THE LOEB CLASSICAL LIBRARY
FOUNDED BY JAMES LOEB

EDITED BY
G. P. GOOLD

PREVIOUS EDITORS
T. E. PAGE   E. CAPPS
W. H. D. ROUSE   L. A. POST
E. H. WARMINGTON

THEOPHRASTUS

DE CAUSIS PLANTARUM

II

LCL 474

THEOPHRASTUS

DE CAUSIS PLANTARUM
BOOKS III–IV

EDITED AND TRANSLATED BY
BENEDICT EINARSON
AND
GEORGE K. K. LINK

HARVARD UNIVERSITY PRESS
CAMBRIDGE, MASSACHUSETTS
LONDON, ENGLAND
1990

Copyright © 1990 by the President and Fellows
of Harvard College
All rights reserved

Library of Congress Cataloging-in-Publication Data

Theophrastus.
De causis plantarum.
(The Loeb classical library: Greek authors; 471, 474, 475)
1. Botany—Pre-Linnean works   I. Title.   II. Series.
QK41.T23   1976   581   76-370781
        ISBN 0–674–99519–8 (v. 1)
        ISBN 0–674–99523–6 (v. 2)
        ISBN 0–674–99524–4 (v. 3)

Typeset by Chiron, Inc.
Printed in Great Britain by St. Edmundsbury Press Ltd,
Bury St. Edmunds, Suffolk, on acid-free paper.
Bound by Hunter & Foulis Ltd, Edinburgh, Scotland.

CONTENTS

SIGLA   vi

DE CAUSIS PLANTARUM
  BOOK III   2
  BOOK IV   190

SIGLA
(see volume I, pages lix–lxi)

U	Vatican City, Urbinas graecus 61 (11th century)
U^d	the diorthotes of U
u	correctors (more probably an Italian corrector) of the 15th century
N	Florence, Laurentian Library, desk 85, 22 (15th century)
v	Venice, Library of St Mark 274 (dated 3 Jan. 1443)
Gaza	Theodorus Gaza's Latin translation (completed 1451)
M	Florence, Laurentian Library, desk 85, 3 (15th century)
C	Oxford, Corpus Christi College 113 (15th century)
H	Harvard College Library 17 (15th century)
a	Aldine Aristotle, fourth volume
P	Paris, National Library 2069 (15th century)
B	Vatican City, Vaticanus graecus 1305 (15th century)
U^c	a correction by the first hand
U^{cc}	such a correcton made in the course of writing
U^{ac}	the reading before correction by the first hand
U^r	a reading due to erasure
U^{ar}	the reading before erasure
U^m	a reading or note in the margin by the first hand
U^t	a reading in the text
U^{ss}	a superscription
U^1	a reading by the first hand

ΠΕΡΙ ΦΥΤΩΝ ΑΙΤΙΩΝ

Γ–Δ

Γ[1]

1.1 ἡ περὶ τῶν φυτῶν θεωρία διττὰς ἔχει τὰς σκέψεις καὶ ἐν δυσίν· μίαν μέν, τὴν ἐν τοῖς αὐτομάτοις γινομένην, ἥπερ ἀρχὴ τῆς φύσεως, ἑτέραν δέ, τὴν[2] ἐκ τῆς ἐπινοίας καὶ παρασκευῆς, ἣν δή φαμεν συνεργεῖν τῇ φύσει πρὸς τὸ τέλος.

ἐπεὶ δὲ περὶ τῆς πρότερον εἴρηται, λεκτέον ὁμοίως καὶ περὶ ταύτης, ὑπὲρ αὐτοῦ τούτου πρῶτον[3] εἰπόντας ὅτι καὶ τῶν δένδρων ἔνια καὶ τῶν ἐλαττόνων ὑλημάτων οὐ δέχεται γεωργίαν· τοῦτο γὰρ ἂν εἴη τῶν μὲν οἷον πέρας, τῶν δὲ ἀρχή, μεταβαίνουσιν ἐκ τῶν αὐτοφυῶν εἰς τὰ διὰ τέχνης.

[1] τὸ γ̄ U^m.
[2] U : τῶν Schneider.
[3] Gaza, Schneider : πρώτου U.

---

[1] Both the nature of the plants (as in Book I) and the environment which is akin to it (as in Book II).

BOOK III

THE EFFECTS OF ART

The study of plants pursues two different investigations in two different fields. The first investigation deals with plants that grow of their own accord, and here the starting-point belongs to their nature[1]; whereas the other starting-point is that which proceeds from human ingenuity and contrivance, which we assert helps their nature to achieve its goal.[2]

A Problem: Plants that Refuse Cultivation

Now that we have discussed the first starting-point,[3] we must similarly speak of the second, first mentioning this very circumstance, that a few trees and smaller woody plants reject cultivation,[4] since this would be the conclusion (so to say) of the one subject and the opening of the next, as we pass from spontaneous occurrences to the results of art.

1.1

---

[2] *Cf. CP* 2 1. 1.
[3] In the whole of Book II.
[4] *Cf. HP* 1 3. 6 (the trees are silver fir, fir, holly; caper and lupine are mentioned among the smaller plants).

THEOPHRASTUS

1.2 ἄτοπον δ' ἂν[1] ἴσως δόξειεν τὸ νῦν λεχθέν· εἰ τὸ προσλαμβάνον θεραπείαν, ὥσπερ χώραν οἰκείαν, μὴ μᾶλλον εὐθενεῖ[2] καὶ καλλικαρπεῖ,[3] συνεργούσης τῇ φύσει τῆς τέχνης. ἀλλ' οὐκ ἔστιν ἄτοπον οὐδ' ἀλλότριον, ἀρχὴν δὲ αὐτοῦ ληπτέον τήνδε· τὴν φύσιν ἑκάστου[4] μὴ ἐκ τῶν αὐτῶν εἶναι, μηδὲ ἐκ μιᾶς τινος ὕλης, μηδ' αὖ πᾶσι ταὐτὸ τέλος τῆς πέψεως καὶ δυνάμεως τῶν καρπῶν, ἀλλ' ἑκάστου πρὸς ὃ πέφυκεν καὶ χυλοῖς καὶ ὀσμαῖς καὶ τοῖς ἄλλοις. μάλιστα δέ πως ἐν δυοῖν τούτοιν ὁρίζομεν τὴν πέψιν· ὀσμαῖς καὶ χυλοῖς, ἀναφέροντες καὶ ταῦτα πρὸς τὴν ἡμετέραν χρείαν (ὥσπερ διωρίσθη καὶ πρότερον).

1.3 ἐπεὶ δ' ἐξ ὧν συνέστηκεν ἕκαστον, καὶ τρέφεται, καὶ τὸ πρόσφορον οὐ κατὰ τὸ ποιὸν μόνον ἐστίν, ἀλλὰ καὶ κατὰ τὸ ποσόν, ἐν ἀμφοτέροις ἂν

---

§ 2 line 6 [Aristotle], *Problems*, xx. 12 (924 a 5–7) [why does caper avoid cultivated land?]: ... περὶ δὴ τούτου καὶ ὅσα ἄλλα τοιαῦτα, δεῖ λαβεῖν ὅτι οὐχ ἅπαντα ἐκ τῆς αὐτῆς ὕλης γίνεται οὐδ' αὔξεται ...

---

[1] U^{css} : U^{ac} omits.
[2] M^r : -εῖν U NM^{ar} P (εὐσθενεῖν H).
[3] M^r : -ειν U (-εῖν u NM^{ar} HP).
[4] τὴν φυσιν εκ- U^c in an erasure.

---

[1] The nature of a tree does best and produces the finest fruit when the tree has its appropriate country: *cf. CP* 1 9. 3 and the other passages adduced at *CP* 3 1. 6.

DE CAUSIS PLANTARUM III

The statement just made might appear odd: that 1.2 a plant that gets tendance besides, just like one that gets its appropriate country,[1] should not do better and bear finer fruit, although art is now cooperating with its nature. But there is nothing odd here or at variance with nature. We would rest our proof on the following considerations: that the nature of each and every plant is not constituted of the same characters,[2] and does not proceed from some matter that is the same for all[3]; and again that all plants do not have the same goal for the concoction and potency[4] of their fruits, but each has instead its own natural goal in point of flavour, odour and the rest.[5] For our part, we[6] somehow determine concoction mainly by these two, odour and flavour, and we moreover refer these to our own uses, according to the distinction drawn earlier.[7]

But since what a plant is formed from it is also 1.3 fed by, and since suitability of food is not only a matter of quality, but also of quantity,[8] the same

---

[2] The formal cause.
[3] The material cause.
[4] "Potency" suggests medicinal use.
[5] The final cause. "The rest" are the "potencies."
[6] The efficient cause (for cultivated plants).
[7] *CP* 1 16. 1.
[8] That is, the food must not only be of the quality that suits the constituents, but the quantities involved must not further some constituents at the expense of others.

εἴη ταὐτὸν[1] οἰκεῖον. ἡ δὲ γεωργία πλῆθός τε τροφῆς καὶ ποιότητα κατασκευάζει, κατεργαζομένη γὰρ ἡ γῆ καὶ ἀπόλαυσιν πλείω δίδωσιν καὶ μεταβάλλει τοὺς χυλούς, ὥστ' εὐλόγως οὐκ ἂν εἴη πρόσφορος ἐνίοις, οἷον ὅσα ξηρά τε καὶ δριμέα καὶ πικρά, καὶ ἁπλῶς ὅσα φαρμακώδη καὶ ἐν τούτοις παρέχεται τὴν χρείαν ἡμῖν· ἐκθηλύνεται γὰρ ἀφαιρουμένων τῶν δυνάμεων καὶ τὰ μὲν ὅλως οὐκ ἐκφέρει καρπούς, τὰ δ' ὑγροτέρους καὶ χείρους, καὶ αὐτὰ ὑδαρέστερα γίνεται, καθάπερ τὸ ἀψίνθιον καὶ τὸ κενταύριον, καὶ ὅλως δὲ πᾶσα φαρμακώδης δύναμις ἢ κατὰ ῥίζαν[2] ἢ κατὰ καρπὸν ἢ κατὰ κλῶνας.

1.4   ἀνὰ λόγον[3] δὲ τούτοις καὶ ὅσα δριμύτητά τινα ἔχειν[4] δεῖ καὶ κατὰ τὴν γεῦσιν, ὧν καὶ ἡ κάππαρις ἔοικεν εἶναι καὶ τὸ σίλφιον καὶ τὸ λάπαθον καὶ ἡ θύμβρα καὶ τὸ θύμον· τὰ μὲν γὰρ ὅλως οὐδὲ

---

§ 3 lines 4–7 ibid. xx. 12 (924 a 16–18): ἡ δὲ γεωργία πέττει καὶ ἐνεργὸν ποιεῖ τὴν τροφήν· ἐξ ἧς συνίστανται οἱ ἥμεροι καρποί.

[1] U : τὸ αὑτοῦ Schneider : τούτοις τὸ Wimmer.
[2] ῥίζαν <ἢ κατὰ καυλὸν> Wimmer.
[3] Wimmer : ἀνάλογον U.
[4] εχειν U : ἔχει u.

DE CAUSIS PLANTARUM III

character in the feeding as in the formation would be appropriate in both quality and quantity. But husbandry brings about both a new quantity of food and a new quality, since tilling the soil not only increases the consumption of a plant but also changes its flavour. Hence it is reasonable that husbandry should fail to suit some plants, namely such as are dry and pungent and bitter, in short all that are medicinal and serve our needs by this or that medicinal character, since the plants are emasculated by the loss of their potencies and some yield no fruit at all, others a more watery and inferior fruit, and the plants themselves become more watery, like wormwood, centaury and in general all drugs and medicines, whether the potency is in root or fruit or twigs.[1]

The results are analogous in plants where a certain pungency is wanted in the taste as well[2]; among these appear to belong caper, silphium, dock, savory and thyme.[3] For on cultivated land some do

1.4

---

[1] *Cf. HP* 9 8. 1: (we must try to speak) "... in general of everything medicinal, such as fruit, flavour-juices, leaves, roots and grass (for the druggists term some of their drugs 'grasses')"; *cf.* also *CP* 6 4. 3.

[2] These are kitchen plants (though a medical use is not expressly denied). Medicinal potencies as such were valued for their physical effects, not for their taste.

[3] *Cf. HP* 1 12. 2 (the fluid in savory and thyme has a certain pungency); *HP* 7 6. 1 (the wild varieties of dock and savory have greater pungency and dryness).

THEOPHRASTUS

φύεται τούτων ἐν τοῖς ἡμέροις ἢ κακῶς, τὰ δὲ χείρω πολλῷ, καθάπερ τὸ λάπαθον καὶ ἡ θύμβρα καὶ ἡ κάππαρις, ἔνια δὲ καὶ τὰς γλυκύτητας αὐτῶν λαμβάνει, καθυγραινόμενα καὶ πληθυόμενα ταῖς τροφαῖς, ὥσπερ τὸ κράνον (καὶ γὰρ τοῦτο χεῖρον ἡμερούμενον γίνεται, καὶ ἁπλῶς δὲ αὐτὸ τὸ δένδρον ὑδαρές τε καὶ μανὸν καὶ ἀσθενές). τῆς δ' οἴης γλυκύτερος μὲν ὁ καρπός, ἧττον δ' εὐώδης,
1.5 ἀφαιρεῖται γὰρ ἡ εὐτροφία[1] καὶ τὸ εὔοσμον. ὁ δ' αὐτὸς λόγος καὶ περὶ σιλφίου καὶ θέρμου[2] καὶ εἴ τι τοιοῦτον ἕτερον. τὸ μὲν γὰρ οὐχ ὁμοίως ἔχει τὴν δριμύτητα διὰ τὸ πλείω καὶ ὑδαρεστέραν εἶναι τὴν τροφήν· ὁ δὲ θέρμος[3] ἄκαρπος γίνεται, καθάπερ ὑλομανῶν καὶ ἐξυβρίζων, ποιεῖ γὰρ καὶ ταύτην τὴν ἐναντίωσιν κατεργασία (καθάπερ εἴρηται), πλῆθος παρέχουσα τροφῆς· ἐκ δὲ τῆς ἀρχῆς ταμιεύεται τὸ σύμμετρον ἕκαστον ἑαυτῷ καὶ λαμβάνει τὴν οἰκείαν χώραν· ἡ γὰρ αὐτοφυὴς γένεσις
1.6 ἐν ταύταις. ἁπλῶς γὰρ τὸ μέγιστον (ὥσπερ πολ-

---

[1] Schneider (⟨διὰ⟩ τὴν εὐτροφίαν Dalecampius) : τὴν εὐτροφιαν U.   [2] u HP : θερμοῦ U N.   [3] u HP : θερμὸς U N.

---

[1] Cf. HP 3 2. 1 (wild trees concoct their fruit less than the cultivated): "All do this with rare exceptions, such as the cornel and the sorb, since they say that here the wild cornel berry and sorb apple are riper and more choice than

not grow at all (or grow poorly), whereas others grow but are far inferior, like dock, savory and caper; a few even get their sweet taste, getting diluted and too full of food, like the cornel berry,[1] for not only does the berry deteriorate on cultivated land, but the tree itself too as a whole gets watery, open in texture, and weak. In the sorb the fruit gets sweeter but less fragrant,[1] since good feeding does away with fragrance too. The story is the same with silphium, lupine and the like.[2] Thus silphium does not have the same pungency under cultivation, since its food is then too abundant and watery. Lupine fails to bear, "running to leaf" (as it were)[3] and getting out of hand. For agriculture works against the nature of the plant in producing this result too (as we said),[4] by filling it with too much food; whereas each plant, left to the government of its own nature, so regulates its food that it gets no more than the right amount, and finds its own appropriate country (since it is in that country that it grows without the aid of man). For in a word the

1.5

1.6

---

the cultivated."

[2] *Cf. HP* 3 2. 1 (continued): "and there are the other cases where a tree or smaller plant will not accept cultivation, like silphium and caper, and among legumes lupine, these being the plants that one would most of all call wild in their nature."

[3] The word was used of the vine; hence the apology "as it were."   [4] *CP* 3 1. 3: "some yield no fruit at all."

THEOPHRASTUS

λάκις εἴρηται) τὸ λαβεῖν οἰκείαν ἀέρα καὶ τόπον · ἐκ τούτων γὰρ ἡ εὐθένεια[1] καὶ εὐκαρπία. ταῦτα δὲ ἐναντία φαίνεται τοῖς παρὰ φύσιν ἡμερουμένοις · εἴς τε γὰρ ἀέρα μαλακώτερον μετατίθενται[2] καὶ τροφὴν ἀλλοιοτέραν λαμβάνουσιν,[3] ἐξ ἀμφοτέρων <δὲ>[4] τούτων ἡ μεταβολὴ καὶ οἷον ἔκστασις τῆς φύσεως. ἐπιμαρτυρεῖ δέ πως καὶ ἐν αὐτοῖς τοῖς γεωργουμένοις τὸ μὴ φιλεῖν ἔνια τὴν ἀκριβῆ καὶ τὴν πολλὴν ἄγαν ἐξεργασίαν, ὥστε καὶ ταύτην ζητεῖν τινα ὅρον · καὶ τρόπον τινὰ οὐδὲν κωλύει κατὰ τὴν μετάβασιν οὕτως ἔνια μηδὲ ζητεῖν ὅλως.

ὅτι μὲν οὖν οὐ πάντα προσδέχεται τὰς γεωργίας φανερὸν ἔστω διὰ τούτων (ἴσως γὰρ τὰ μὲν ὅλως οὐ δεῖται, τῶν δ' οὐκ ἐξευρίσκομεν τὰς αἰτίας).

2.1 ἡ δὲ γεωργία, δυοῖν ὄντων ἐξ ὧν αἱ τροφαὶ καὶ εὐθένειαι,[5] τῆς γε[6] γῆς[7] παρασκευάζει βοήθειαν,

---

[1] U : εὐσθένεια u.
[2] u (μετατεθέντα Schneider) : μετατιθέντα U.
[3] U : λαμβάνοντα Schneider.   [4] HP.
[5] U N : εὐσθένειαι u HP.   [6] ego : τε U (Heinsius omits).
[7] γῆς <καὶ τοῦ ἀέρος · τῆς γῆς μόνον> Schneider : γῆς <καὶ τοῦ ἀέρος, τὴν ἀπὸ τῆς γῆς> Wimmer.

most important thing (as we have often said)[1] is that a plant should get the air and the locality that are appropriate to its nature, since from these come well-being and good bearing. But air and locality are seen to work against plants that are subjected to cultivation against their nature, for the plants are removed to air that is too gentle for them and get food that is of too different a character, and from both of these circumstances comes the change and departure (as it were) from their nature. This view receives a certain support from the fact that even among plants that are under cultivation some do not like the care to be over-precise or over-lavish, and so care too requires a certain restriction; and in a way there is nothing to prevent that by gradual increase in the restriction there should finally be some plants that want no care at all.

We shall take it, then, as made clear by this discussion why not all plants accept cultivation. For perhaps whereas some plants will have none of it at all, others reject it for reasons that we have not succeeded in discovering.

Agriculture: Aims and Procedures

Of the two sources[2] of food and well-being, agriculture can at least remedy conditions in the 2.1

---

[1] *HP* 2 2. 8; 2 5. 7; 3 3. 2. *CP* 1 9. 3; 2 7. 1; 2 16. 7; 2 19. 6.   [2] The appropriate air and locality (*CP* 3 1. 6).

THEOPHRASTUS

τὸν ἀέρα γὰρ οὐκ ἐπ' αὐτῇ ποιόν τινα ποιεῖν, ἀλλὰ δεῖ πρὸς τὰς μεταβολὰς καὶ τὰς ὥρας τὰς γινομένας αὐτὰ τὰ δένδρα καὶ τὴν χώραν διακεῖσθαί πως, ἵνα τε ἀπολάβῃ[1] τὴν προσαύξησιν καὶ τροφὴν[2] καὶ μὴ πάσχῃ μηδὲν ὑπὸ τῶν ἐναντίων (δεῖ γὰρ καὶ πρὸς ταῦτα φυλακῆς), λέγω δ' οἷον τομάς τε καὶ διακαθάρσεις καὶ βλαστολογίας καὶ κοπρίσεις καὶ σκαπάνην[3] καὶ ὅσα ἄλλα πραγματεύονται (καὶ περὶ τῆς ἀροσίμου δὲ καὶ σπερματουμένης ὁμοίως).

2.2

τοῖς μὲν γὰρ οὐ ταὐτά, τοῖς δ' οὐκ ἴσα, τοῖς δ' οὐ τὸν αὐτὸν τρόπον, οὐδὲ τὴν αὐτὴν ὥραν ἀποδιδόναι δεῖ · τοῖς δὲ ἴσως οὐδὲ ποιεῖν ἔνια τὸ ὅλον. ἃ δὴ καὶ φαίνονται διαιροῦντες οἱ γεωργοί, καθάπερ καὶ ἐν τῇ διακαθάρσει τὰ μὲν σιδήροις, τὰ δὲ ταῖς χερσίν, τὰ δ' ἀγκύραις τισὶν κελεύοντες ἀφαιρεῖν, οἷον τῆς τε ἐλάας καὶ τῆς ἀπίου καὶ τῆς μηλέας,

---

[1] u : ἀπολαύει U (-η N HP).
[2] U : τῆς πρὸς τὴν (Wimmer omits τὴν) αὔξησιν τροφῆς Schneider, Wimmer.
[3] Heinsius : σκεπάνην U.

---

[1] As opposed to orchard land and vineyards; trees (including the vine) were propagated by cutting and dig-

DE CAUSIS PLANTARUM III

ground, since it is not in its power to change the character of the air. Instead the trees themselves and the land must meet the changes of the air and the course taken by the seasons by being put into a certain state, so that the trees may obtain their due of growth and food and escape the effects of adverse circumstances (protection against these being also needed): I mean such measures as pruning, trimming, thinning the fruiting shoots, and manuring, spading and the other operations about which the agriculturists elaborate (so too with ploughed and seeded land).[1]

2.2

Thus with some plants one must not apply (1) the same kinds of things[2]; with others (2) the measurements involved must not be the same; with others the procedure must not be applied (3) in the same manner or (4) at the same season; and with others perhaps (5) there are some things that must not be done at all.[3] These are the distinctions that the agriculturists are in fact observed to make: for instance (3) in pruning they recommend the removal to be carried out in some cases with iron tools, in others with the bare hands, in others with what they call "anchors,"[4] as with olive, pear and apple,

---

ging holes. Land was ploughed for cereals (and some grains and vegetables); for many it was merely seeded.

[2] So of manure (*CP* 3 9. 5).

[3] *Cf.* Plato, *Phaedrus*, 268 B 6–8, 272 A 6–7.

[4] Apparently a pruning-hook shaped like an anchor.

THEOPHRASTUS

ὅπως μὴ ἑλκούμενα πονῇ διὰ τὴν λεπτότητα καὶ ξηρότητα τῶν κλάδων · καὶ γὰρ ἀφαιρεῖν δεῖ τούτων ταῦτά[1] τε καὶ τὰ τρώξανα μόνον. ὁμοίως δὲ καὶ τὰ τῶν ἀμπέλων · διαιρετέον τάς τε πρωτοτόμους[2] καὶ τὰς ὀψιτόμους, καὶ τὰς βραχυτόμους καὶ τὰς μακροτόμους, καὶ τοῖς ἄλλοις ὡσαύτως, ἀναφέροντας ἀεὶ πρὸς τὸ τέλος.

2.3

ὁ μὲν οὖν σκοπὸς οὗτος, καὶ ἡ καθ' ἕκαστα διαίρεσις.

ὑπὲρ ἑκάστου δέ ἐστιν ὁ λόγος ὁ τὴν αἰτίαν ἔχων, ἣν δεῖ μὴ λανθάνειν · ὁ γὰρ ἄνευ ταύτης ποιῶν καὶ τῷ ἔθει καὶ τοῖς συμβαίνουσιν κατακολουθῶν κατορθοῖ μὲν ἴσως, οὐκ οἶδεν δέ (καθάπερ ἐν ἰατρικῇ) · τὸ δὲ τέλειον ἐξ ἀμφοῖν. ὅσοι δὲ καὶ τὸ θεωρεῖν μᾶλλον ἀγαπῶσιν, αὐτὸ τοῦτο ἴδιον τοῦ λόγου[3] καὶ τῆς αἰτίας.

---

[1] U : τὰ αὐά Schneider.   [2] U : πρωϊτόμους?
[3] U^c : τῶι λόγωι (?) U^{ac}.

---

[1] *Cf.* Plato, *Phaedrus*, 271 B 1–5.
[2] *CP* 3 2. 1.   [3] *CP* 3 2. 2–3.
[4] *Cf.* Plato, *Phaedrus*, 268 A 8–C 4, 270 B 4–7 (for the example of medicine as an art and as mere experience);

DE CAUSIS PLANTARUM III

to avoid any suffering of the trees from wounds, owing to the thinness and dryness of their twigs (indeed in these trees only the twigs and deadwood should be removed). So too with the procedures for pruning the vine: we must distinguish between (4) vines pruned early and vines pruned late, vines (2) pruned long and vines pruned short, and so with the rest of these matters (1, 5), always relating the distinction of treatment to the end in view.[1]

2.3

This[2] then is the aim of agriculture, and these[3] are the distinctions in its various procedures.

Agriculture: Theories and Disputes

About each procedure there is the account that gives its reason, and the reason must not escape us. For the man who carries out the procedure in ignorance of the reason, guided by habit and by the event, may perhaps succeed, but he does not *know* (just as in medicine), and complete possession of the art comes from both.[4] As for those who have a greater love for understanding,[5] this very thing, understanding, comes only when we have the account and the reason.

---

269 C 2, D 2, 272 A 2 (for completeness as involving theoretical knowledge); 270 B 4–5, D 1–7, 271 B 1–2, D 4–6 (for dividing the subject dealt with and the procedures used); 271 B 2–4, D 6–7 (for teaching that gives the reasons).   [5] Philosophers.

THEOPHRASTUS

2.4 ἀντιλέγεται δὲ[1] περὶ πολλῶν, καὶ τὰ μὲν ἁπλῶς, τὰ δ' εἰς τὸ βέλτιον καὶ χεῖρον, ὥσπερ κατὰ τὰς ἄλλας τέχνας. ἐνταῦθα δὲ καὶ ἰδιώτερόν τι συμβαίνει· πρὸς γὰρ τὴν ἑαυτῶν ἔνιοι χώραν τετραμμένοι τὰ πρόσφορα πολλάκις καθόλου λέγουσιν. ὁτὲ δὲ καὶ ἀμφοτέρως γινομένων ἄκριτόν ἐστι τὸ βέλτιον· ὁποῖα γὰρ ἂν ᾖ τὰ ἀπὸ τοῦ ἀέρος[2] συμβαίνοντα, τοιαῦτα ἀποβαίνει καὶ[3]

2.5 κατὰ τὰς ἐργασίας. οὐδὲ γὰρ δεῖ[4] μᾶλλον ἀκολουθεῖν[5] τῇ τοῦ ὅλου καταστάσει καὶ περιφορᾷ τῶν δένδρων καὶ φυτῶν καὶ σπερμάτων,[6] ὡς πολλάκις ἁμαρτανόμενα[7] τῇ ἀπὸ τούτων εὐκρασίᾳ[8] καὶ[9] τῇ αὐτῶν[10] δυνάμει τὰ μὲν ὑπομένει, τὰ δέ τινα ἀναμάχεται, καθάπερ καὶ τῶν ἀνθρώπων φύσις[11] <τὰ>[12] ὑπὸ τῆς ἰατρικῆς. ὥστε αἱ μὲν ἀντιλογίαι πρὸς ταῖς κοιναῖς τῶν τεχνῶν καὶ διὰ ταύτας γίνονται τὰς αἰτίας.

[1] U : γὰρ Schneider.
[2] τοῦ ἀέρος Schneider (aeris Gaza : ἀέρος Itali) : θέρους U.
[3] καὶ <τὰ> H^c P.
[4] u : δὴ U.   [5] u HP : -εῖ U N.
[6] τῶν ... σπερμάτων U : *quam plantarum: ac seminum naturam* Gaza : <ἢ τῇ> τῶν ... σπερμάτων <φύσει> Basle ed. of 1541 : <ἢ τῇ φύσει> τῶν ... σπερμάτων Schneider.
[7] <τὰ> ἁμαρτανόμενα Scaliger.
[8] U : ἀκρ- N HP.

DE CAUSIS PLANTARUM III

Many matters are in dispute; in some the dispute is a simple question of yes or no, in others it is a question of better or worse, just as in the other arts. But in agriculture something rather special also occurs: some experts have their view fixed on their own country and often state as a general rule what is successful there. Sometimes moreover both rules work and there is no deciding which is the better, since the outcome of the agricultural measures depends on accidents of weather. For one should not in fact be governed by the celestial conditions and revolution[1] rather than by the trees and slips and seeds, since owing to good tempering contributed by these conditions and to their own power they often either resist mistaken measures or in some cases recuperate, just as a human constitution does with the mistakes of medicine. And so disputes in addition to those common to the other arts arise in agriculture for these reasons.

2.4

2.5

---

[1] That is, the farmers' calendars, which reckoned by the position of the sun and the rising or setting of various stars and constellations, foretelling the weather by counting from these in days (the "revolution") and advising what procedures to undertake at what time.

---

[9] [καὶ] Schneider.
[10] U : αὐτῶν Scaliger.
[11] <ἡ> φύσις HP.   [12] Schneider.

THEOPHRASTUS

οὐ μὴν ἀλλὰ καὶ συνομολογεῖται πολλάκις[1] παρὰ πάντων, ὥσπερ ἐξ αὐτῶν τῶν πραγμάτων εἰληφότα πίστιν, τὰ μὲν καθόλου, τὰ δὲ διαιρούμενα γένεσι[2] καὶ χώραις, οἷον εὐθὺς τὰ περὶ γενέσεις καὶ φυτείας, ὑπὲρ ὧν οὐ χαλεπὸν εἰπεῖν τὰς αἰτίας.

2.6 ἀεὶ γὰρ δεῖ φυτεύειν καὶ σπείρειν εἰς ὀργῶσαν τὴν γῆν· οὕτω γὰρ καὶ ἡ βλάστησις καλλίστη, καθάπερ τοῖς ζῴοις ὅταν εἰς βουλομένην πέσῃ τὸ σπέρμα τὴν ὑστέραν. ὀργᾷ δ' ὅταν ἔνικμος ᾖ καὶ θερμὴ καὶ τὰ τοῦ ἀέρος ἔχῃ σύμμετρα· τότε γὰρ εὐδιάχυτός τε καὶ εὐβλαστής, καὶ ὅλως εὐτραφής ἐστιν. τοῦτο δ' ἐν δυοῖν ὥραιν γίνεται μάλιστα τοῖς γε δένδροις, ἔαρι καὶ μετοπώρῳ, καθ' ἃς καὶ φυτεύουσι μᾶλλον, καὶ κοινοτέρως ἐν τῷ ἦρι· τότε γὰρ ἥ τε γῆ δίυγρος καὶ ὁ ἥλιος θερμαίνων ἄγει καὶ ὁ ἀὴρ μαλακός ἐστιν καὶ ἐρσώδης, ὥστε ἐξ ἁπάντων εἶναι τὴν ἐκτροφὴν καὶ τὴν εὐβλαστίαν.

---

[1] U : *multa* Gaza : πολλὰ καὶ Schneider.
[2] Schneider : γενέσει U.

---

[1] "Planting" (*phyteúō*) is used of putting a slip (*phytón*) in the ground, "sowing" (*speírō*) of sowing seeds, especially

DE CAUSIS PLANTARUM III

Nevertheless there is often agreement on all sides, as if the rules had been accredited by the facts themselves, some in a general formulation, and some drawing distinctions between the kinds of plant and kinds of country, such as to begin with the rules about propagation and planting, for which it is not hard to give the reasons.

Planting: The Seasons

So one must always plant and sow[1] when the earth is in heat, since then the sprout that comes forth is best, just as in animals when the seed enters a womb desiring it.[2] The earth is in heat when it is moist and warm and the weather temperate, since then it is loose and sends the shoot up quickly and is in general nutritious. This receptivity occurs chiefly, at least for trees, at two seasons, spring and autumn, these being in fact the times when more planting is done, and for a greater number of different trees it is done in spring, for the earth is then soaked through and the sun by its warmth brings about growth and the air is mild and dewy, so that all this combines to rear the slips and make them sprout well.

2.6

---

of grain (*spérmata*).

[2] *Cf.* Aristotle, *On the Generation of Animals*, ii. 4 (739 a 31–35); Plutarch, *Quaestiones Convivales*, v. 7. 3 (681 F).

THEOPHRASTUS

2.7 ἔχει δέ τινα καὶ τὸ μετόπωρον τοιαύτην κρᾶσιν (διὸ καὶ τὰς ἐκβλαστήσεις[1] ἔφαμεν γίνεσθαι τῶν δένδρων). οἱ δὲ ἐπαινοῦντες αὐτὴν μᾶλλον τοῦ ἔαρος[2] τοιαῦτα λέγουσιν· ὅτι τὰ φυτὰ θερμῆς οὔσης ἔτι[3] τῆς γῆς ῥιζοῦται κάλλιον, δεῖ[4] δ' ἀεὶ τὴν ἀρχὴν ἰσχυροτέραν ποιεῖν ἀφ' ἧς καὶ ἡ τῶν ἄλλων γένεσις καὶ ὅλως ἡ ζωή· διὸ καὶ τὴν βλάστησιν ἅμα τῷ ἦρι καλὴν γίνεσθαι καὶ ἀθρόον, ἐκβεβηκυίας ἤδη καὶ κυούσης ταύτης,[5] τοῦ δὲ ἦρος εἰς ψυχρὰν τιθεμένων τὴν γῆν (ψυχρὰν γὰρ ἔτι διαμένειν ἐκ τοῦ χειμῶνος) ῥιγοῦν[6] καὶ κακοβλα-
2.8 στεῖς γίνεσθαι τὰς ῥίζας. καὶ διὰ τοῦτο πάντα κελεύουσι τοῦ μετοπώρου μᾶλλον φυτεύειν, ὅσα δύναται βλαστάνειν (ἔνια γὰρ οὐ δύναται, καθάπερ ἄπιος καὶ μηλέα, καὶ ὅλως τὰ λεπτὰ καὶ ξυλώδη· καὶ γὰρ ξηρὰ[7] φύσει, καὶ διὰ τὴν ἀσθένειαν οὐχ ὑπομένει τοὺς χειμῶνας), τῷ[8] τὸ μὲν μετοπωρινὸν[9] ἔγκυμον εἶναι καὶ ἐπίφορον, τὸ δὲ ἐαρινὸν ἄρτι κυΐσκεσθαι· τὸ δὲ τίκτειν ἀμφοτέροις

---

[1] U (*cf. HP* 3 5. 5) : ἐπι- Schneider.
[2] u : ἐτέρος U : ἀέρος N HP.
[3] Gaza (*adhuc tepido*), Scaliger : θέρμης οὔσης ἐπι U.
[4] U : δεῖν Gaza (*oporteat*), Schneider.
[5] ταύτης ego (τῆς ῥίζης Schneider) : τῆς γῆς U.
[6] With ῥιγοῦν H breaks off; it lacks the rest of the *CP*.

DE CAUSIS PLANTARUM III

Autumn too has a somewhat similar tempering (which is why we said[1] that trees have sproutings then). Those who favour autumn over spring reason as follows: "Slips root better when the ground is still warm (and we should always make the beginning[2] stronger, from which the other parts are produced and from which indeed the tree lives). This is why at the coming of spring the upper parts come out fine and all at once, the beginning having by then grown out and being pregnant with them. In spring on the other hand the slips are set in earth that is cold (since it remains cold from winter) and are chilled and the roots come out poorly." For this reason they recommend planting in autumn rather than in spring all trees that can grow when so planted (for some are unable to do so, like pear, apple and in general all that are thin and woody, since they are both naturally dry and also are too weak[3] to stand the winter), because the slip planted in autumn is pregnant and near its term, whereas the one planted in spring is just conceiving; but both

2.7

2.8

---

[1] *CP* 1 12. 3; *cf. CP* 1 6. 3.
[2] That is, the root.
[3] Because they are thin.

---

[7] u : ξηρᾷ U.
[8] τῷ ego (ἔτι Gaza [*item*], Itali : ἔτι δὲ Basle ed. of 1541) : ὅτι U.
[9] Gaza (*plantam autumnalem*), Itali : μετόπωρον U.

τὴν αὐτὴν ὥραν, ὥστε συμβαίνειν[1] τῷ μὲν ἄρτι κυϊσκομένῳ πολλὰ τυφλὰ τίκτειν τῶν βλαστημάτων, τῷ δὲ πάλαι κύοντι καὶ ἐπιφόρῳ τὰ πλεῖστα τέλεα καὶ καλά. σχεδὸν γὰρ ταύτας καὶ τοιαύτας λέγουσιν τὰς αἰτίας.

3.1 οὐ μὴν ἀλλὰ (καθάπερ εἴρηται) κοινοτέρα πᾶσιν ἢ τοῖς πλείστοις καὶ τοῖς σπέρμασι[2] καὶ φυτοῖς ἡ τοῦ ἔαρός[3] ἐστιν· αὐτά τε γὰρ τὰ φυτὰ προωρμημένα, καὶ <ὁ> ἀὴρ[4] μαλακὸς καὶ εὐβλαστής, καὶ ὅλως ἡ ὥρα γονιμωτάτη· καὶ αἱ ἡμέραι θερμότεραι καὶ μῆκος ἔχουσαι ταχείας ποιοῦσι τὰς βλαστήσεις. ἡ δὲ ῥίζωσις ἰσχυρὰ καὶ οὕτω διά τε <τὴν>[5] τῶν φυτευτηρίων ὁρμὴν (εἰς ἄμφω γὰρ ὁμοίως ἐστί) καὶ διὰ τὸ τὴν γῆν[6] ὀργᾶν ἢ[7] καὶ ἔτι τὸν ἀέρα συνεκτρέφειν (ὁ γὰρ ἥλιος οὐ μόνον δοκεῖ τὰ ὑπὲρ γῆς ἀλλὰ καὶ τὰ ὑπὸ γῆς εὐτραφέστερα καὶ καλλίω ποιεῖν· σημεῖον δ', ὅτι τῶν ῥιζῶν ἡ βλάστησις ἢ οὐ γίνεται πορρωτέρω τῆς τοῦ ἡλίου δυνάμεως ἢ χείρων).

---

[1] Gaza (*evenire*), Schneider : -ει U.
[2] Schneider (*fructuum* Gaza) : ὅποις U.
[3] Gaza, Italì : ἀέρος U.
[4] Schneider (ἀὴρ u) : ἀὴρ U.
[5] u.   [6] U^c : ὀργὴν U^{ac}.
[7] [ἢ] Gaza, Schneider.

bring forth at the same season, with the result that the tree which is just conceiving brings forth many of its shoots blind,[1] whereas the one that is well along in its pregnancy and near its term brings forth most of them perfect and fine. These (one may say) and the like are the reasons that they cite.

Nevertheless (as we said)[2] planting in spring is the one more common to all or at least most seeds as well as slips. For the slips themselves are then well on the way to sprouting, the weather is mild and good for sprouting and in general the season is best at generation; and the increasing warmth and length of the days make the sprouting rapid. And the rooting is strong at this season too, not only because the slips are astir with the impulse to grow (for the impulse is equally in both directions, down as well as up), but also because the earth is in heat, or because, in addition to this, the air helps to rear the plant (the sun being considered not only to make the parts above ground plumper and finer but also the parts below, the proof being that the roots either do not grow beyond the reach of the sun's power[3] or else grow more poorly).

3.1

---

[1] *Cf.* Aristotle, *On the Generation of Animals*, ii. 4 (739 a 31–35); Plutarch, *Quaestiones Convivales*, v. 7. 3 (681 F).

[2] *CP* 3 2. 6 (where however seeds were not included).

[3] *Cf. CP* 1 12. 7 and *HP* 1 7. 1 (of roots): "... none goes deeper than the sun reaches, since it is heat that generates them."

THEOPHRASTUS

3.2 ὁ δὲ χειμὼν ἐπιλαμβάνων[1] τὴν φυτείαν ἔνιά γε δοκεῖ φθείρειν, οἷον τῆς μὲν ἐλάας ἀφιστὰς τὸν φλοιόν, τῆς δὲ συκῆς παχύνων καὶ πηγνὺς τὸν ὀπὸν τὸν ἐν τῇ κράδῃ,[2] καὶ τῶν ἄλλων σχεδὸν τῶν πλείστων τοιοῦτόν τι πονούντων.[3] ἀλλὰ γὰρ ταῦτα μὲν ἴσως ἀφορίζοιτ' ἂν ὁ φάσκων τὴν μετοπωρινὴν εἶναι τὴν βελτίω, τοῖς δεχομένοις λέγων.

ὅτι <δὲ>[4] πλείω ταῦτα <καὶ>[5] τῆς φύσεως αὐτῆς ἐστι σημεῖα τῶν πρὸς τὴν φυτείαν ὡρῶν τό τε παορμᾶν[6] αὐτὰ τὰ δένδρα τοῦ ἦρος καὶ τὸ κοινοτέραν εἶναι πᾶσι ταύτην, δῆλον ὡς φυσικώτε-
3.3 ρον[7] ἄν τις θείη ταύτην. ἐπεὶ καὶ ὅπου περὶ Κύνα καὶ τοὺς ἐτησίας εὐθενεῖ καὶ ἐπιβλαστάνει τὰ δένδρα, τηνικαῦτα καὶ τὰς φυτείας ποιοῦνται, πολλῷ[8] δέον ἀκολουθεῖν τῇ τοῦ ἀέρος κράσει καὶ τῇ τῶν φυτῶν ὁρμῇ· τάχα δὲ καὶ ἡ ὁρμὴ γίνεται διὰ τὸ περιέχον. ὅπου δ' αὖ θερινὸς ὄμβρος πολύς,

---

[1] u : ὑπο- U (ὑπο- N aP).
[2] Schneider : καρδίᾳ U.   [3] Schneider : ποιούντων U.
[4] ὅτι δὲ ego (Verum quum Gaza : εἰ δὲ Wimmer) : ετι U (ἔτι u) : ἔστι N : ἔτι δὲ aP.   [5] ego.
[6] U : προορμᾶν Schneider (cf. προωρμημένα CP 3 3. 1).
[7] U : φυσικωτέραν Schneider.
[8] πολλῷ u (cf. CP 6 11. 10) : πολῶν U.

---

[1] Cf. CP 3 2. 8.
[2] Cf. the calendar (Introduction, vol. I, p. xlix f.: July 18:

DE CAUSIS PLANTARUM III

When winter follows the planting it is held to destroy at least some of the slips, separating for example the bark from the olive slip and thickening and congealing the juice in the branch of the fig; and one might say that most of the rest labour from similar ill effects. But the advocate of autumn planting would perhaps exclude these slips and say that it is better for the slips that are receptive to it.[1]

3.2

But because there are more slips that suffer, and because the impulse of the trees themselves in the spring and the greater universality of spring planting in all plants are indications that this season for planting belongs to the very nature of the plants, one would evidently account planting in spring the more natural procedure. Indeed in regions where even in the dog days and when the Etesians blow[2] trees do well and have a second sprouting growers also do their planting at that very time, so important is it to be guided by the temperate state of the air and the impulse in the slips; in fact the impulse perhaps is due to the surrounding air. Again in regions where there is heavy sum-

3.3

Sirius rises in the morning (Eudoxus); the Etesians blow for 55 days ... August 26: the Etesians cease (Callippus). *Cf.* Aristotle, *Meteorologica*, ii. 5 (361 b 35–363 a 2): "The Etesians blow after the summer solstice and the rising of Sirius, and neither when the sun is closest nor yet when the sun is far; and they blow by day and stop at night."

ὥσπερ ἐν Αἰθιοπίᾳ καὶ ἐν Ἰνδοῖς, ἢ περὶ Αἴγυπτον ὁ Νεῖλος, ἐνταῦθα δὴ πρὸ τούτων ἢ μετὰ τούτους εἰκὸς τὴν φυτείαν ἁρμόττειν· τηνικαῦτα γὰρ ἡ τοῦ ἀέρος κρᾶσις σύμμετρος.

3.4 εἴη δ᾽ ἂν διελεῖν καὶ τοῖς κατὰ φύσιν τόποις πρὸς τὰς[1] ὥρας, οἷον τοὺς μὲν εὐκράτους τῷ ἀέρι τοῦ ἦρος, τοὺς δὲ ῥοώδεις καὶ ἐπόμβρους καὶ ἑλείους θέρους ὑπὸ τὸ ἄστρον (ὥσπερ καὶ ἐν τῇ Λακωνικῇ πολλοὶ[2] φυτεύουσιν), τοὺς δ᾽ αὐχμώδεις μετοπώρου· συμβήσεται γὰρ οὕτω, θερμῆς οὔσης ἐν βάθει τῆς γῆς κατὰ χειμῶνα, ψυχροῦ δὲ τοῦ πέριξ, τὴν αὔξησιν κατακλειομένην εἰς τὰς ῥίζας ἰέναι· πλείονος δ᾽ ὄντος καὶ ἰσχυροτέρου τοῦ ῥιζώματος, πλείων[3] ἡ βλάστησις ἔσται καὶ καλλίων.

καὶ περὶ μὲν ὡρῶν τῶν εἰς τὰς φυτείας ἱκανῶς εἰρήσθω.

4.1 ἐπεὶ δὲ ὑπόκειται τὴν γῆν ἔνικμόν τε δεῖν καὶ

[1] καὶ τοῖς ... τὰς U : κατὰ φύσιν καὶ τοῖς τόποις τὰς πρὸς <τὴν φυτείαν> Schneider.
[2] u : πολὺ U : πολλὰ N aP.   [3] u : πλεῖον U.

[1] Cf. Aristotle, *Meteorologica*, i. 12 (349 a 4–7): "Further in Arabia and Ethiopia the rains come in summer and not in winter, and come in downpours, often several times in the same day..."

mer rain, as in Ethiopia[1] and India[2] (or in Egypt the Nile),[3] it is presumable that a good time for planting is before or after the rains, this being the moment when the air is tempered to the right degree of wetness.

One could also distinguish the seasons for planting by the regions naturally appropriate to a season, as follows: one should plant regions with well-tempered air in spring; regions full of flowing water, or rainy, or marshy, in summer during the dog days (this is the practice of many growers in Sparta); and regions with a scarcity of water in autumn, for the result will be that since in winter the ground deep below the surface is warm, whereas the surrounding air is cold, growth will be shut in by the cold and pass to the roots, and greater and stronger rooting will lead to greater and finer sprouting above ground.

3.4

Let this suffice for the treatment of the seasons for planting.

Planting: Preparation of the Ground

We have laid it down[4] that the earth should be

4.1

---

[2] The monsoon.
[3] *Cf.* Aristotle, Fragments 246–248 (ed. Rose[3]) [the inundation of the Nile occurs in summer and is due to the summer rains in Ethiopia].
[4] *CP* 3 2. 6.

THEOPHRASTUS

εὐδίοδον εἶναι ταῖς ῥίζαις, ὅπως εὐμήκεις κα⟨ὶ⟩ παχεῖαι¹ καὶ ἰσχυραὶ γίνωνται, διὰ ταῦτα δεῖ τοὺς [τε]² γύρους προορύττειν ἐκ πολλοῦ,³ μάλιστα δὲ ἐνιαυτῷ πρότερον, ὅπως ἡ γῆ καὶ ἡλιωθῇ καὶ χειμασθῇ καθ' ἑκατέραν τὴν ὥραν· ἄμφω⁴ γὰρ ταῦτα ποιεῖ μανὴν καὶ χαύνην. ἔνιοι δὲ καὶ διορίζουσι τοὺς χρόνους ἀπὸ τροπῶν θερινῶν μέχρι Ἀρκτούρου·⁵ τότε γὰρ τήν τε γῆν διαχεῖσθαι⁶ μάλιστα, καὶ τὴν ἔξω καὶ τὴν ἐν τοῖς γύροις,⁷ καὶ βλαστάνειν ὅλως οὐδέν. φυτεύουσι δὲ τὴν μετοπωρινὴν φυτείαν μετὰ Πλειάδος δύσιν, δεξάμενοι τὸ⁸ ἐπὶ τῷ ἄστρῳ ὕδωρ, ὅπως ἔνικμος ἡ γῆ γενομένη παρέχῃ τροφήν.

4.2 καὶ τοὺς γύρους οὐκ εὐθὺς συμπληροῦσιν, ὅπως ῥιζωθῇ τὰ κάτω πρότερον· εἰ δὲ μή, φέρονται πρὸς τὰ⁹ ἄνω (τρέφει γὰρ καὶ αὔξει πάνθ' ὁ ἥλιος καὶ ὁ ἀήρ). ὄντων δὲ τῶν μὲν¹⁰ βαθυρρίζων, τῶν δ' ἐπιπολαιορρίζων, διὰ τοῦτο τοὺς γύρους οὐκ

---

¹ Gaza, Basle edition of 1541 : ταχεῖαι U.
² Schneider.
³ u : πολὺ U : πολλὰ N aP.
⁴ N aP : ἄφω U.
⁵ Uᶜ from ἀκ-.
⁶ Gaza, Itali : διακεῖσθαι U.
⁷ u : γυροῖς U.

moist and provide an easy passage for the roots, to allow them to become long, thick and strong. The holes must therefore be dug a good time beforehand,[1] preferably a year, to expose the earth to the sun in summer and to the cold in winter. For both exposures make it open in texture and loose. Some experts even specify the time for digging: from the summer solstice to the rising of Arcturus, since the earth (they say) is then loosest both outside and inside the holes, and nothing sprouts at all.[2] Autumn planting is carried out after the setting of the Pleiades, first waiting for the rain that falls at this time,[3] to let the earth become moist and provide food.

The holes moreover are not filled up immediately 4.2 after planting, to let the lower part of the slip strike root first; otherwise the roots move higher (since sun and air nurture everything and make it grow).[4] Trees being distinguished into the deep and shallow-rooted, the holes intended for the shallow-

---

[1] *Cf. HP* 2 5. 1 (on planting): "They recommend ... digging the holes as long as possible beforehand ..."

[2] The hole is thus easy to dig and needs no weeding.

[3] *Cf. CP* 3 23. 1, where Cleidemus says that rain falls on the seventh day after the setting of the Pleiades.

[4] The sun provides heat, the air rain.

---

[8] Schneider : τε U.   [9] U : τὸ Wimmer.
[10] τῶν μὲν U^cm (with index) : U^t omits.

THEOPHRASTUS

ἰσοβαθεῖς ὀρύττουσιν τοῖς ἐπιπολαιορρίζοις (οἷον ἐλάᾳ καὶ συκῇ[1]), βουλόμενοι πιέζεσθαι[2] καὶ ὥσπερ ἀντιταττόμενοι πρὸς τὰς φύσεις.

4.3 ὅπως δὲ καὶ τῶν ὑδάτων τοῦ χειμῶνος ἀπολαύωσιν[3] καὶ τοῦ θέρους καταψύχωνται (δεῖ γὰρ δὴ πρὸς ἀμφοτέρας τὰς ὥρας παρεσκεύασθαι), διὰ τοῦθ' ὑποβάλλουσι κάτω λίθους, ὅπως συρροὴ γίνηται τοῦ ὕδατος, καὶ τοῦ θέρους οὗτοι καταψύχωσι τὰς ῥίζας. οἱ δὲ κληματίδας ὑποτιθέασιν, οἱ δὲ κέραμον[4] παρακατορύττουσιν ὕδατος, οἱ δὲ ξύλον κνημοπαχές,[5] εἶτ' ἐξαιροῦσιν, ὅπως ἔχῃ τροφὴν ἀεὶ τὰ φυτά, δικμαζομένης τῆς γῆς καὶ συρρεόντων τῶν ὑδάτων. ἁπλῶς γὰρ τοῦτο δεῖ τηρεῖν· ὅπως καὶ πρὸς τοὺς ὄμβρους τοὺς γινομένους, καὶ πρὸς τὸν ἀέρα καὶ τὸν ἥλιον, ἕξει συμμέτρως· αἱ γὰρ τροφαὶ καὶ αἱ αὐξήσεις διὰ τούτων.

4.4 ἐπεὶ δὲ ἡ γῆ βορείοις μὲν πεπηγυῖα καὶ ξηρά, νοτίοις δὲ κεχυμένη καὶ ἔνικμος, ὡσαύτως δὲ καὶ τὸ φυτὸν ὑγρότερον καὶ αὐτὸ ἑαυτοῦ μᾶλλον, διὰ

[1] a (-ῇ P) : συκαι U : συκαὶ u N.
[2] Scaliger : πιεσθαι U.
[3] ἀπολαύωσι u : ἀπολαύουσιν U.
[4] U : κεράμιον Schneider.
[5] Wimmer (κἂν κνημοπαχὲς Schneider) : καν ημοπαχες U.

DE CAUSIS PLANTARUM III

rooted kind (as olive and fig) are not dug only as deep as the roots would ordinarily go,[1] since the agriculturists wish the slip to be under pressure to send its roots deeper, and set themselves (as it were) against its nature.

To let the roots profit from the rains in winter and be cooled in summer (since provision should be made for both seasons), stones are placed at the bottom to allow rain water to collect and to keep the roots cool in summer. Others put vine branches at the bottom, others bury a pot of water alongside, and others a stick, which they later remove, as thick as a man's leg, all so that the slips may be ensured a constant supply of food, the earth being well moistened and the rain water collecting. We must take care, in a word, that there shall be adequate provision for dealing with the rains that fall and with the air and the sun, these being the means whereby the young trees feed and grow.

4.3

When northerly winds prevail the earth is stiffened and dry, but when southerly ones prevail it is loose and moist, and the slip is similarly more springy and truer to type.[2] It is therefore better to

4.4

---

[1] *Cf. HP* 2 5. 1 (on planting): "They recommend ... digging the holes always deeper than the roots, even for shallow-rooted trees."

[2] Literally "more its own self"; for the expression compare *CP* 3 7. 10, 4 3. 4.

THEOPHRASTUS

ταῦτα βελτίων ἡ τοῖς νοτίοις φυτεία· ταχεῖα γὰρ ἡ ῥίζωσις καὶ ἡ βλάστησις ὅταν ὀργῶν εἰς ὀργῶσαν τεθῇ, καὶ τὰ τοῦ ἀέρος ᾖ μαλακὰ καὶ εὐμενῆ (δεῖ γὰρ δὴ τὸ μέλλον ἔσεσθαι καλὸν ἅμα κάτωθεν καὶ ἄνωθεν βλαστάνειν). τοῖς δὲ βορείοις ἅπαντα τἀναντία γίνεται, ῥιγοῦν γὰρ καὶ κακοπαθεῖν τὸ φυτόν· ἔτι δέ, τῆς γῆς πεπηγυίας, οὔτε ῥιζοῦσθαι δύναιντ' ἂν ὁμοίως οὔτε βλαστάνειν.

ἡ μὲν οὖν τοιαύτη παρατήρησις τοῦ ἀέρος ἂν εἴη.

5.1 τὰ δὲ φυτευτήρια δεῖ λαμβάνειν ἀπὸ νέων τε[1] τῶν δένδρων (ἢ ἀκμαζόντων) καὶ ὅλως λειότατα καὶ ἰθύτατα καὶ ὡς πάχιστα·[2] καὶ γὰρ ἀντιλαμβάνεται καὶ ἰσχύει τὰ τοιαῦτα μάλιστα, καὶ διὰ τὴν ἡλικίαν καὶ τὸ πάχος εὐβλαστότερα τυγχάνει· τό τε[3] γὰρ λεῖον ὥσπερ ὑγιὲς καὶ ἀπήρωτον, τὸ δὲ τραχὺ καὶ ὠζωμένον,[4] ἄλλως[5] τε καὶ τυφλοῖς ὄζοις, ὥσπερ πεπηρωμένον, ἐάν τε ζῶσιν,

[1] U : γε Wimmer. [2] ego : κάλιστα U.
[3] τότε U : τὸ μὲν Schneider.
[4] Gaza (*nodatum*), Heinsius : ὀζόμενον U.
[5] N aP (ἄ- u) : ὅλως U.

[1] *Cf. HP* 2 5. 1: "They recommend taking the slips as stout (reading πάχιστα for ταχιστα U) as possible."

plant during southerly winds, rooting and sprouting being rapid when slip and ground are both in heat and the weather is gentle and propitious (for to make a fine tree the slip must sprout above ground at the same time as it roots below). When northerlies blow, on the other hand, the opposite of all this occurs, the slips getting chilled and suffering hardship; moreover, since the ground is stiff with cold, the consequence is a lessening in their ability both to root and sprout.

Here, then, is a rule to follow in the matter of the air.

Planting: The Slips

The slips should be taken from trees that are 5.1 young or in their prime, and should in any case be the smoothest and straightest and as stout as possible,[1] since not only does such a slip take hold best and have most strength, but its youth and stoutness make it also sprout better. For the smooth slip is as it were sound and unmaimed, whereas the slip that is rough and full of nodes,[2] especially if the nodes are blind,[3] is as it were maimed,[4] and if the nodes

---

[2] The notion includes the buds and twigs coming from the nodes.

[3] That is, if the buds or twigs coming from the nodes bear no fruit.

[4] "Maimed" ($\pi\eta\rho\acute{o}\varsigma$ and its derivatives) is often used of blindness.

THEOPHRASTUS

ἐκβεβλαστηκὸς[1] εἰς τούτους ἀσθενέστερον, ὅσων[2] μηδ' ἡ φύσις τοιαύτη, καθάπερ τῶν κλημάτων. διὰ τοῦτο γὰρ καὶ ἐπὶ τῶν δένδρων ἔνιοι μοσχεύουσιν, οἱ δὲ περιαιροῦσιν τὴν θάλλειαν[3] τῶν κλάδων,[4] ὅπως μὴ ἐξαναλώσῃ[5] τὴν δύναμιν εἰς τὴν βλάστησιν. ἔτι δ' εὐθύτης λειότης τε[6] εὔρουν καὶ εὐδίοδον ποιεῖ τὴν ῥοήν, ὥστε ταχείας εἶναι τὰς αὐξήσεις.

5.2

ὀρθῶς δὲ καὶ τὸ μᾶλλον ἐξ ὁμοίας γῆς, εἰ δὲ μή, [μὴ][7] χείρονος λαμβάνειν· ἡ μὲν γὰρ οὐδεμίαν ποιήσει μεταβολήν, ἡ δ' ἐπὶ τὸ βέλτιον εὐτροφοῦντος·[8] μέγα δὲ αἱ μεταβολαὶ τοῖς ἀσθενέσιν, ἀσθενὲς δὲ τὸ φυτόν. διὰ ταῦτα γὰρ καὶ τὰς θέσεις τῶν φυτευμάτων τὰς αὐτὰς ἀποδιδόασιν κατὰ τὰ πρόσβορρα καὶ νότια, καὶ πρὸς ἕω καὶ δυσμάς, ὡς ἐπὶ τῶν δένδρων εἶχε, βουλόμενοι τηρεῖν ὅτι[9] μάλιστα καὶ μηδὲν τῆς φύσεως καὶ τῶν εἰωθότων μετακινεῖν, ὡς οὐκ ἂν ῥᾳδίως ἐνεγκόν-

---

[1] u : ἐκβεβλαστικῶς U : ⟨τὸ⟩ ἐκβ. Schneider (*quod . . . ediderit* Gaza). [2] Gaza, Schneider : ὅσον U. [3] ego : θάλειαν U.
[4] U^r N aP : κελάδων U^{ar}. [5] u : -ει U.
[6] Schneider : λειότητος U. [7] Cagnatus.
[8] ego (εὔτροφος οὖσα Wimmer) : συντροφοῦντος U.
[9] Gaza (*quam*), Itali : ἔτι U.

---

[1] *Cf. HP* 2 6. 3, *CP* 3 11. 5 (a layer has roots stronger than those sent down by a slip).

DE CAUSIS PLANTARUM III

live, a slip that has let its growth pass into them is weaker (unless its nature is to do so, as with the branches of the vine). For the reason why some experts even layer the slips on the trees,[1] and others strip the olive cutting of its leaves, is to prevent it from exhausting its power on growth above ground. Again, straightness and smoothness make the flow of food plentiful and direct, so that the slips grow fast.

5.2

Another good recommendation is to take the slip preferably from similar soil, or failing that, from poorer,[2] since the new soil in the first case will involve no change, in the second, a change for the better, the slip getting plenty of food; and changes are grave matters for whatever is weak, and the slip is weak. For this is why the recommendation is given that the slip should face in the same direction—north, south, east and west—as it did on the tree,[3] the intention being to disturb so far as possible nothing that belongs to its nature or to the circumstances to which it is accustomed, in the belief that the slips would not easily support a

---

[2] This passage is referred to at *CP* 3 24. 1. *Cf. HP* 2 5. 1: "They recommend taking the slips ... from land similar to the one you are going to plant them in or from worse."

[3] *Cf. HP* 2 5. 3 (recommendations about slips): "Some recommend that also the lower part of the slips that have roots should be bent under, and be placed in the same position that some of the trees had, those facing north and east and south."

35

THEOPHRASTUS

τῶν μεταβολήν.

5.3 ἐπεὶ καὶ τοὺς τόπους ὅτι μάλιστα ὁμοίους ζητοῦσι διὰ τὰς αὐτὰς αἰτίας, καὶ προμοσχεύοντες[1] φυτεύουσιν· ἰσχυρότερα γὰρ καὶ ὥσπερ ἤδη βεβιωκότα. καὶ τὰ φυτὰ μάλιστα μὲν ὑπόρριζα λαμβάνουσιν, ἔχει γὰρ εὐθὺς καὶ ἀρχάς· εἰ δὲ μή, μᾶλλον ἀπὸ τῶν κάτω ἢ τῶν ἄνω, καὶ γὰρ ταῦτα ἐμβιώτερά ἐστιν, πλὴν ἀμπέλου καὶ συκῆς καὶ εἴ τι ἄλλο ὑγρόν (ὥσπερ εἴρηται). τὰ γὰρ ὑγρὰ κάλλιον[2] ἀπὸ τῶν ἄνω βλαστάνει, τὰ δ' ἄλλα[3] διὰ τὴν ξηρότητα καὶ τὴν λεπτοδερμίαν τὰ μὲν ὅλως οὐ βλαστάνει, τὰ δὲ χεῖρον. τὸ δ' ὑπόρριζον,[4] ὡς προπεπονημένον[5] τι τῆς φύσεως ἔχει, καθάπερ τὸ

5.4 μεμοσχευμένον, ᾧ δεῖ χρῆσθαι. καὶ διὰ τοῦτο βέλτιον λέγουσιν ὀρθὰ κελεύοντες τὰ τοιαῦτα τιθέναι, καὶ μὴ ὑποβάλλοντάς[6] τι μέρος (ὥσπερ τοῖς ἀρρί-

---

[1] u : πρὸσμ- U.   [2] U^c : -ιω U^{ac}.
[3] δ' ἄλλα U^{cc} : δε χειρον U^{ac} (skipping some 58 letters).
[4] Wimmer (ὑπόρριζα Schneider): ὑπέρριζων U (-ον u): ὑπὲρ ῥίζων N : ὑπὲρ ῥιζῶν aP.   [5] u aP : -μένον N : -μένων U.
[6] ego : ὑποβάλλόντές (sic) U.

---

[1] "First making them grow roots" renders *promoscheúontes*. *Móschos* is literally a "calf," and as a calf to a cow, so a young tree with roots of its own stands to the parent. The "moschos" when planted has roots, either produced by

change.

Indeed for the same reason experts look for regions as similar as possible and plant the slips after first making them grow roots,[1] since they are stronger then and have as it were already established independent life. And they prefer to take a slip with some root attached, since then it starts with a beginning.[2] Otherwise they prefer slips from the lower part to those from higher up, these too being better at taking, except for the vine[3] and fig and other fluid trees (as we said),[4] since fluid trees sprout better from slips taken from their upper parts, whereas the rest are so dry and thin-skinned that they either do not sprout at all when so planted or sprout worse. The slip with some root attached is preferred since it has a part of its nature already prepared, like the slip that has been made to grow roots, and of this we must take advantage. This is why it is better advice to plant such slips straight, and not lay flat a certain portion of them (as is done

5.3

5.4

layering or after separation from the parent, and is distinct from a slip with some root attached. This is taken directly from the tree, where it grew as a sucker or as a side-shoot from the base of the trunk.

[2] *Cf. CP* 3 2. 7, note 2.

[3] *Cf. HP* 2 5. 3: "One must take the slips with some root attached if possible; otherwise from the lower rather than from the upper parts of the tree, except for the vine . . ."

[4] *CP* 1 1. 4; 1 3. 1.

THEOPHRASTUS

ζοις), ὅπως ἡ ῥίζα βλαστάνῃ · [1] συμβαίνειν [2] γὰρ δοκεῖ πλαγίων τιθεμένων ἀφαυαίνεσθαι τὰς πρότερον, ἄτοπον δὲ ζητεῖν ἑτέρας ἀποβάλλοντα τὰς ὑπαρχούσας. ὀρθῶς δὲ καὶ τὸ μὴ πολὺ τῆς γῆς ὑπερέχειν τὸ φυτόν, εἰ δὲ μή, γίνεται δυσαυξές, ὅταν ᾖ πλέον τοῦ τρεφομένου τὸ πονοῦν · ἔνια δὲ καὶ ἀφαυαίνεται, καθάπερ ἡ συκῆ καὶ εἴ τι ἄλλο μανόν · ἅμα δὲ καὶ ἀθροωτέρα φέρεται μᾶλλον ἡ τροφὴ πρὸς τὴν βλάστησιν, ὃ καὶ ἐπὶ τῶν κατα-

5.5 κοπτομένων συμβαίνει. τὰ γὰρ παρὰ τὴν γῆν ὅτι μάλιστα κοπέντα θᾶττον παραγίνεται τῶν ἐν ὕψει · φανερὸν δὲ καὶ ἐπὶ τῶν τῆς ἀμπέλου φυτῶν καὶ εἴ τι ἄλλο τομὴν ζητεῖ κατὰ τὴν φυτείαν. τῶν δὲ τῆς ἐλάας καὶ τῶν μυρρίνων, καὶ ὅλως ὅσα φύεται [3] μείζω, πάντων ἀποστέγουσι [4] τὰς τομάς, ὅπως μήθ᾽ ἥλιος μήθ᾽ ὕδωρ λάβῃ · κίνδυνος γὰρ νοσῆσαι ῥαγέν · περιαλείφουσιν δὲ οἱ μὲν πηλὸν μόνον, οἱ δὲ σκίλλαν ὑποτιθέντες, εἶτ᾽ ἄνωθεν τὸν πηλόν, ἐπὶ τούτῳ δὲ τὸ ὄστρακον · δοκεῖ γὰρ ἡ μὲν σκίλλα χλωρὸν παρέχειν, ὁ δὲ πηλὸς ἐκείνην

---

[1] βλαστάνῃ u : -ει U.   [2] u : -ει U.
[3] U : φυτεύεται (?) Schneider.
[4] Gaza (*contegunt*) : ἀποστέγουσα U.

---

[1] *Cf. HP* 2 5. 3 (recommendations about slips): "One

DE CAUSIS PLANTARUM III

with the rootless slips) to make the root come out[1]; since the earlier roots are believed to wither away when the lower portion of the slip is planted at an angle, and it is folly to seek for new roots by throwing away the ones you have. It is also good advice not to let the slip project far above the ground, since otherwise its growth is slow, the part exposed to hardships being larger than the part that feeds, and some such slips even wither away, as with the fig and other trees of open texture. Then too when the slips do not project far the food passes on with a more concerted movement to bring about sprouting, and the same results when the slips are pruned short. For slips cut back as close as possible to the ground come out faster than those cut back higher, and this is evident not only in slips of the vine but in all others that require to be cut back when planted. In the slips of olive, myrtle and in general all slips of larger growth, the cut is sealed off to keep sun and rain from getting at it, since otherwise there is danger that it will split and get diseased. Some experts merely smear the cut with mud, but others first cover the cut with a squill, then cover the squill with mud and the mud with a shard, since the squill is considered to keep the stump fresh, the mud to

5.5

should plant the ones with roots straight, and should lay flat about a handsbreadth or a little more of those that are rootless. Some recommend laying flat the same length of slips with root attached too . . ."

THEOPHRASTUS

τηρεῖν, τὸ δ' ὄστρακον τὸν πηλόν.

6.1 ἡ δὲ κόπρος, ὅτι μὲν καὶ μανοῖ τὴν γῆν καὶ διαθερμαίνει, δι' ὧν ἀμφοτέρων ἡ εὐβλαστία, φανερόν· ὑπὲρ δὲ τῆς χρήσεως διαμφισβητοῦσιν, καὶ οὐχ ὡσαύτως χρῶνται πάντες, ἀλλ' οἱ μὲν εὐθὺς ἀναμίξαντες τῇ γῇ πρὸς τὸ φυτὸν προσβάλλουσιν, οἱ δ' ἀνὰ μέσον ποιοῦσι τῆς τε πρώτης γῆς καὶ τῆς ἐπάνω, κάτωθεν γὰρ παραλαμβανομένην[1] εἰς τὸ ἄνω φέρεσθαί φασιν ὅταν ὕσῃ· βέλτιστον γὰρ εἶναι τὸν χυλόν, ἄνωθεν δ' ἐξικμάζεσθαι ὑπὸ τοῦ
6.2 ἡλίου, καὶ ὕοντος οὐ διικνεῖσθαι τὸ[2] κάτω. πάντες δὲ τό γε τοσοῦτον συμφωνοῦσιν, ὥστε μὴ δριμεῖαν καὶ ἰσχυράν, ἀλλὰ κούφην, διὸ καὶ μάλιστα χρῶνται τῇ τῶν λοφούρων· ἡ γὰρ δριμεῖα καὶ ἰσχυρὰ διαθερμαίνει μᾶλλον, ἡ δὲ καὶ ξηραίνει.

χρὴ δὲ καὶ πρὸς τὴν χώραν ἑκάστην ποιεῖν τὸ ἁρμόττον· οἷον ἐὰν μέν τις <ἐν>[3] ἐμπύρῳ

---

[1] U : παραβαλλομένην Schneider.
[2] U : εἰς τὰ Schneider.   [3] Schneider.

---

[1] *Cf. HP* 2 7. 4 (of manure): "... some plants require it to be pungent, some to be less so, and some require it to be quite light. The most pungent is that of man. So Chartodras says that the best is that of man, second of swine, third of goat, fourth of sheep, fifth of ox and sixth of pack-

DE CAUSIS PLANTARUM   III

guard the squill and the shard the mud.

Planting: Manure

It is evident that manure not only gives the earth 6.1
a loose texture but also warms it through, both of
which lead to rapid sprouting. But there is a
dispute about how it is to be used, and not all the
experts apply it in the same way. Some mix it
directly with the earth and put the mixture around
the slip. Others put the manure in a layer between
the earth that makes the bottom of the hole and the
fill of earth at the top, since if put below it moves
up after a rain (the liquid solution that moves up
being, they say, the best part), but if put on top it
loses its moisture to the sun, and when it rains the
liquid does not reach the bottom. All however agree 6.2
to this extent: that the manure should not be
pungent and strong but light, and this is why they
chiefly employ that of pack-animals,[1] since pungent
and strong manure heats too thoroughly, and some
of it also dries the slips.

Planting: Suiting the Country

We must also do what is suited to the various
kinds of country. So if one is setting slips in a torrid

---

animals. Litter manure is different and rates differently,
some being weaker than the last, some better."

THEOPHRASTUS

[φυτῶι]¹ φυτεύῃ, τοὺς γύρους ὕδατος ἐμπιπλάναι πρότριτα, καὶ ἐπειδὰν² ἀναπίωσι,³ τότε βάλλειν,⁴ ὅπως ἔνικμος ἡ γῆ γενομένη μᾶλλον δέχηται καὶ καλλίω ποιῇ τὴν ῥίζωσιν.

6.3 ἐὰν δὲ ἐν ἁλμώδει ᾖ ἐφάλμῳ, λίθους περιτιθέναι περὶ τὸ πρέμνον τοῦ φυτευτηρίου καὶ περιχωννύναι γῆν, ὅπως ἀποστέγωσι τὴν ἁλμυρίδα. συμφέρει δὲ καὶ ἄμμον πότιμον παραβάλλειν καὶ ψήφους ἐκ ποταμοῦ ἢ χαράδρας· ἀποστέγει γὰρ καὶ ταῦτα καὶ ἅμα δι' αὐτῶν παρέχεταί τινα δύναμιν.

ἐὰν δὲ ἔφυδρα καὶ ναματώδη,⁵ τάφρους ὀρύσσοντα, τὰς μὲν πλαγίους ἵνα τὸ ὕδωρ δέχωνται, τὰς δὲ ὀρθίας, καὶ λίθων πληροῦντα καὶ γῆς, ὥστε μὴ ἅπτεσθαι τῆς σκαπάνης, εἶτα ἄμμον ἐμβάλ-
6.4 λοντα⁶ καὶ χοῦν. ἅπαντα γὰρ ταῦτα ὠφελεῖ πρὸς τὴν ὑπερβολήν· αἵ τε γὰρ πλάγιοι τάφροι, δεχόμεναι τὸ ὕδωρ, ξηροτέραν ποιοῦσιν, αἵ τ' ὄρθιαι,⁷ κάτω λίθους ἔχουσαι, δέχονται τὴν συρροήν· ἔτι⁸

---

¹ Scaliger : *solo* Gaza, τόπῳ Heinsius.
² aP : ἐπι (ἐπὶ u N) τὴν U.
³ aP (-ιν u) : ἀναπιῶσι U (-ιῶσι N).
⁴ τότε βάλλειν U : *serere* Gaza : καταβάλλειν Wimmer.

country one must fill the holes with water two days before and when the holes have absorbed it insert the slips, so that the earth, now full of moisture, may be more ready to accept them and make them root better.

If the soil is brackish or has a coating of brine we 6.3 must circle the base of the slip with stones and cover all this with earth so that the stones may shut out the brine. It also helps to put in brine-free sand or pebbles taken from a river or a torrent bed, since the sand and pebbles also shut out the brine and have besides a certain power of their own.[1]

If there is much surface water and many streams one must dig ditches, some of them with sloping banks to catch the water and others with steep banks, and fill the steep ditches with stones and earth to a level not reached by hoeing, and then cover the fill with sand and loose soil. For all these 6.4 measures help against the excess of water. For the sloping ditches catch the water and make the ground drier, and the steep ditches with stones at the bottom receive the water that is caught, and

---

[1] They cool the roots (*cf. CP* 1 18. 1; 3 4. 3); the stones are too far from the roots to do this.

---

[5] U : δ' ἐν ἐφύδρῳ καὶ ναματώδει Schneider after Gaza.
[6] aP : -αι (no accent U) u N.
[7] u : ορθριαι U (ὄ- N aP).
[8] Schneider : ὅτι U.

THEOPHRASTUS

δ' ἡ ἄμμος καὶ ὁ χοῦς ἀναξηραίνουσι.[1] διὰ τοῦτ' οὐδὲ [ἂν] οἵ γε <ἂν> ὦσι[2] λίθοι ταῖς τοιαύταις οὐκ ἐκλεκτέοι, ὑπάρχει γὰρ φύσει τὸ βοηθοῦν· ὅλως ἐν ὁποιᾳοῦν, ἐὰν ὀρύττων τις λίθους εὕρῃ ἢ τρόχμαλον ἢ ἄμμον ἢ γῆν μοχθηράν, οὐ κακῶς[3] τὸ μὴ συμμειγνύναι μηδὲ σκεδαννύναι ταύτην, ἀλλὰ τοσούτῳ βαθύτερον τὸν γῦρον ἢ τὴν τάφρον ὀρύξαντα τόν τε τρόχμαλον ὑποστρωννύναι καὶ

6.5 τὴν ἄμμον. ὁ γὰρ λίθος ὁ μὲν πλατὺς καὶ συμφυὴς καὶ[4] βλάπτει τὰ δένδρα· ὁ δὲ τρόχμαλος ὑποκάτω τεθεὶς ψῦχός τε παρέχεται τό θ' ὕδωρ δέχεται, καὶ ταῖς ῥίζαις εὐδίοδον ποιεῖ τὸν τόπον. ἁπλῶς δὲ τούς γε λίθους ἐκλέγειν οὐδ' ἐκ γῆς ψιλῆς οἴονταί τινες <συμ>φέρειν·[5] καὶ γὰρ ἀλέαν παρέχειν τοῦ χειμῶνος καὶ ψῦχος τοῦ θέρους.

ὁμοίως δὲ καὶ κατὰ τὰς ἄλλας διαφορὰς τῆς χώρας τὰ πρόσφορα ληπτέον. ὡς γὰρ ὅλως εἰπεῖν οὐδενὸς ἔλαττον, ἀλλὰ πάντων πρῶτον καὶ μέγιστον, τὸ δύνασθαι θεωρῆσαι ποῖον ἐν ποίᾳ χώρᾳ φυτευτέον, οὐ μόνον ἁπλῶς, ἀλλὰ καὶ αὐτῆς τῆς

---

[1] u N^r (-σιν N^{ar}) aP : αναξηραίνουσαι U.
[2] ego (οὐδ' ἐάν που ἐνῶσι Schneider) : οὐδὲ ἂν οιγεωσι U.
[3] U : οὐκ ἄκος u.
[4] ego : ἡ U : ἢ u N : aP omit.
[5] Heinsius.

DE CAUSIS PLANTARUM III

furthermore the sand and loose soil of the fill dry out the ground. For this reason one should not even remove any stones there are in such land, since the remedy in this case is naturally provided, and in general, whatever the character of the land, if in digging one comes upon stones or rubble or sand or poor soil, it is no bad rule not to mix this in with the earth or scatter it on the top, but to dig the hole or the ditch that much deeper and spread the rubble or sand on the bottom. For when stone is flat and unbroken it actually injures the trees, whereas in the form of rubble and placed below it provides coolness, catches the water, and makes the ground permeable to the roots. In short some experts believe it inadvisable to remove the stones even from a grain field, pointing out that they provide warmth in winter[1] and coolness in summer.

6.5

Similarly too with the other distinctions of country: we must find the proper measures with the distinctions in view. For to speak generally, the point that is second to none, but first and most important of all, is this: to be able to see what tree to plant in what country,[2] not only in the country as a whole, but also in what part of a continuous country itself

---

[1] *Cf. CP* 3 20. 5.
[2] *Cf.* Plato, *Phaedrus*, 271 D 7–E 2; also *HP* 2 5. 7: "Most important is the assigning of the suitable country to each tree, since then the tree does best."

συνεχοῦς, ὅταν ἀνωμαλὴς ᾖ κατὰ τοὺς τόπους
6.6 ἑκάστους. ὥσπερ γὰρ λέγεται πολλάκις, ἡ οἰκεία
μεγίστην ἔχει¹ ῥοπὴν καὶ πρὸς ἀντίληψιν καὶ
πρὸς εὐκαρπίαν, ὃ καὶ καθ᾽ ὅλων τῶν γενῶν ἐστιν
καὶ ἐν αὐτοῖς τοῖς ὁμοειδέσιν. οἷον ὡς μὲν ἁπλῶς
εἰπεῖν ἡ συκῆ φιλεῖ τοὺς ξηροὺς τόπους, οἱ γὰρ
ὑγροὶ σήπουσιν ἢ οὐ καλῶς πεπαίνουσιν, διὰ τὸ
καὶ αὐτὸν εἶναι τὸν καρπὸν ὑγρόν. οὐ μὴν ἀλλ᾽
ὅσαι² γε ὑδρεύεσθαι ζητοῦσιν, καθάπερ ἡ Λακωνι-
κή, τὰ ἔφυδρα³ ζητοῦσιν. ὡσαύτως δὲ καὶ τῶν
ἀμπέλων αἱ μὲν στερεαὶ καὶ πυκναί, καθάπερ καὶ
πρότερον ἐλέχθη, τὴν ὀρεινὴν μᾶλλον φιλοῦσιν, αἱ
δὲ μαναὶ καὶ ὑγραὶ τὴν πεδεινήν· ἑκατέραις γὰρ
ἡ τροφὴ πρὸς τὴν φύσιν σύμμετρος, ταῖς μὲν
ἐλάττων οὖσα, ταῖς δὲ πλείων.⁴

6.7 ὅλως δὲ τὰ ἀκρόδρυα καλὰ δοκεῖ περὶ τὰς ὑπω-
ρείας γίνεσθαι (σημεῖα δὲ ποιοῦνται τὴν αὐτομά-
την γένεσιν· ὅπου γὰρ ἡ φύσις αὐτὴ γεννᾷ, τοῦτον οἰκειότατον εἶναι τόπον). αἱ δὲ καθ᾽ ἕκαστα

---

¹ P : ἔχειν U N a.   ² Schneider : ὅσοι U.
³ u : ὕφυδρα U (ὔ- N aP).   ⁴ u aP : πλειω U (πλείω N).

---

¹ *Cf. CP* 3 1. 6 with the passages there cited.
² *Cf. HP* 2 5. 7: "Among the trees of the same class too one must not be ignorant of the ones (*sc.* the countries)

when the country varies from one district to
another. For (as we repeat)[1] the appropriate country has the greatest weight both in determining that
a tree takes hold and that it bears well, and this
applies not only to whole classes but also the
varieties within the same class.[2] For example the
fig as a class likes dry regions, since wet ones rot the
fruit or fail to ripen it properly, because the fruit is
itself full of fluid. Nevertheless the varieties, such
as the Laconian,[3] that like watering like regions
where there is ground water. So too with the vine:
the solid and close-textured varieties (as we said
earlier)[4] prefer mountain country, the open-textured and fluid ones prefer the plain, since each
of the two varieties thus gets the right amount of
food for its nature, the amount being less for the
former, greater for the latter.

6.6

Fruit trees as a class are considered to turn out
fine in the foothills,[5] the proof adduced being their
growing there of their own accord, since wherever
the tree's own nature generates it unaided, this
(they say) is the locality most appropriate for the

6.7

appropriate to them."

[3] *Cf. HP* 2 7. 1: "The fig when watered has finer leafage but its fruit is poorer, except for the Laconian fig, and this loves water."   [4] *CP* 2 4. 7.

[5] *Cf. HP* 2 5. 7: "They say that for olive, fig and vine as a class the plain is most appropriate, for fruit trees the foothills."

THEOPHRASTUS

διαφοραὶ δῆλον ὅτι κυριώταται, καὶ γὰρ ὑπώρειαι πολλαὶ καὶ παντοῖαι · καὶ τὸ ὅλον, ὥσπερ μία τις αὕτη διαφορά, καθ' ὕψος καὶ ταπεινότητα, πολλῶν οὐσῶν ἑτέρων καὶ μειζόνων (ὥσπερ ἐν ταῖς ἱστορίαις εἴπομεν), ὧν οὐδεμίαν ἀθέατον[1] εἶναι χρή.

6.8 τὴν γοῦν λειμωνίαν καὶ ἔφυδρον σχεδὸν οἱ πλείους ὁμολογοῦσιν ἀγαθὴν εἶναι ταῖς ἀμπέλοις, ὥσπερ καὶ τὴν εὔγειον[2] ἐλάαις καὶ συκαῖς · αἱ μὲν γὰρ ὑγροτέρας δέονται καὶ μαλακωτέρας τροφῆς, αἱ δὲ ξηροτέρας καὶ σωματωδεστέρας · ὅλως δ' αἰεὶ[3] τὰ μὲν ξηρὰ ξηρὰν ζητεῖ χώραν, τὰ δὲ ὑγρὰ ὑγράν.

ὡς δ' ἁπλῶς εἰπεῖν ἀρίστη πᾶσιν ἡ μανὴ καὶ κούφη καὶ ἔνικμος, εὐτρεφὴς[4] γὰρ μάλιστα καὶ εὐαξής (εἰ μὴ ὅσα διὰ τὴν ἰσχὺν λαμβάνοντα πλῆθος τροφῆς ἐξυβρίζει, καθάπερ ἡ ἀμυγδαλῆ · ταύτῃ γὰρ ἡ λεπτόγειος οἰκειοτέρα,[5] καὶ τόπος

---

[1] ego (cf. θεώμενον Plato, Phaedrus 271 D 8 : ἀθεώρητον Coray) : θετέον U.
[2] u : εὐγείων U.
[3] U^{cc} (αι from ε).   [4] N (-ῆς U) : εὐτραφὴς u aP.
[5] Scaliger : ὑγροτέρα U.

---

[1] HP 3 2. 5: "Nevertheless on the great mountains, such as Parnassus, Cyllene, the Pierian and Mysian Olympus

DE CAUSIS PLANTARUM III

tree. But the particular distinctions are evidently what matters most, since there are also many widely different kinds of foothills. Indeed on a general view this is itself but one of the distinctions, resting on the degree of elevation, among many other more important distinctions (as we said in the History),[1] none of which must escape our study.

So the majority (one might say) of experts agree 6.8 that meadow land and land with surface water is good for the vine,[2] as rich land is good for olive and fig; since the vine requires food with more fluid and softness, the others food with more dryness and body. In geneal, dry trees always like the country dry, whereas fluid ones like it fluid.

Put simply, the best land for all trees is loose, light and damp, since it feeds them best and makes them grow rapidly, except for trees so vigorous that they take too much food and get out of hand, like the almond.[3] For thin soil is the more appropriate for

---

and the like elsewhere, all wild trees grow because of the great variety of the localities, the mountains having both regions with lakes and water and dry regions, both regions of soil and regions of rocks, and the meadows that lie in between, and one may say all the varieties of land that there are; furthermore they have some regions that are in basins and have fair weather, others that are elevated and wind-swept, so that the mountains are able to produce all types of trees, including those that are found in the plains." [2] *Cf. CP* 2 4. 4.

[3] *Cf. CP* 2 16. 8, 3 18. 2.

εὐδιεινὸς καὶ ἥλιος·[1] ἐναντίως δὲ τῇ Εὐβοϊκῇ,[2] παλίσκιος καὶ δροσερός).

6.9 οὐ μικρὸν δὲ οὐδὲ πρὸς τὰ πνεύματα καλῶς κεῖσθαι τοὺς τόπους, ὥστε καὶ τοῦτο θεωρητέον.

εὔπνουν μὲν γὰρ ἅπαντα ζητεῖ, τροφαί τε γὰρ ἐν τοῖς τοιούτοις βελτίους καὶ οἱ καρποὶ πεπειρότεροι. πνευματώδη δὲ καὶ προσήνεμον οὐδὲν ὡς εἰπεῖν τῶν γε ἡμέρων, ἀναυξῆ γὰρ καὶ μικρὰ καὶ καυλώδη[3] γίνεται διὰ τὰς πληγὰς τὰς ὑπὸ τῶν πνευμάτων. ἔνια δὲ οὐδὲ τῶν ἀγρίων· οὐ γὰρ τελειοκαρπεῖ, καθάπερ οὐδὲ τὴν Εὐβοϊκήν φασιν, ἀλλὰ μέχρι τοῦ ἄνθους ἀφικνεῖσθαι, τὰς δ' ἐν τοῖς ὑπηνέμοις εὐαξεῖς τε καὶ πολυκάρπους γίνεσθαι.

τὸ γὰρ ὅλον (ὥσπερ πολλάκις εἴρηται) συμμετρίας τινὸς ἔοικεν ἕκαστα δεῖσθαι πρὸς τὴν φύσιν, ὁμοίως ἔν τε ταῖς ἄλλαις τροφαῖς καὶ ταῖς τοῦ ἀέρος μεταβολαῖς. ἀλλὰ τοῦτο μὲν ἂν εἴη καθόλου καὶ κοινόν.

7.1 τὰς δὲ μανότητας καὶ πυκνότητας τῶν φυτευο-

---

[1] U : εὔηλος (εὔειλος Schneider) Itali.
[2] Scaliger : εὐβοῆι U.
[3] U^c : καυκαλώδη U^{ac}.

DE CAUSIS PLANTARUM III

the almond and a region with clear weather and plenty of sun; the sweet chestnut on the contrary likes a shady spot with dew.

Nor is it of small importance that the land should be well placed with regard to the winds. So this point is another that must be studied. 6.9

All trees like a well ventilated region, since there the food is better and the fruit mellower. But no tree (one might say), at least no cultivated tree, likes a region that is windy and to windward, since the buffeting they get from the winds makes them stunted, small, and thickset. Even some wild trees do not like such a region, since they fail to ripen their fruit (so they say that the sweet chestnut does not like such a region either, and that it gets no further than the blossom, whereas on the leeward slope it grows tall and bears abundantly).

Each tree, in a word, appears to require a certain quantitative adjustment to its nature (as we have often said),[1] not only in the matter of food but in that of the changes of the air. But this would be a general point and of common application everywhere.

Planting: Spacing

The spacing of the slips must be considered not so 7.1

---

[1] *CP* 1 10. 5 (adjustment to the season), *CP* 2 3. 4 (to the atmosphere), *CP* 2 9. 13 (of food and air).

μένων οὐχ οὕτω πρὸς τὸν τόπον ὡς πρὸς αὐτὰ τὰ
φυτευόμενα σκεπτέον — ὅσα τε φιλόσκια καὶ μή,
καὶ ὅσα μακρόρριζα καὶ βραχύρριζα· τὰ μὲν γὰρ
δῆλον ὅτι μανά, τὰ δὲ πυκνὰ φυτευτέον. φιλόσκια
δὲ ὧν οἱ καρποὶ ξηρότατοι[1] καὶ πυρηνώδεις
(ὥσπερ ῥόας καὶ μυρρίνου), καὶ ὅσα φύσει μανά τε
καὶ ξηρὰ καὶ μὴ μακρόρριζα (καθάπερ ἡ δάφνη).

7.2 τῶν μὲν γὰρ ὁ καρπὸς ἡλιούμενος στρυφνὸς γίνεται, τὰ δὲ ὥσπερ προβολῆς δεῖται καὶ πρὸς τοὺς
χειμῶνας καὶ πρὸς τὰ καύματα, διὰ τὴν ἀσθέ-
νειαν· οὐκ ἐνοχλεῖ γὰρ τῷ ταρρῷ διὰ τὴν βραχυρ-
ριζίαν. καίτοι τινὲς ἅπαντα κελεύουσιν διὰ πολλοῦ φυτεύειν, ὅπως μὴ σύνταρρα γίνηται, μηδ᾽
ἀναυξῆ, τροφήν τε ἐλάττω λαμβάνοντα καὶ τοῦ
πνεύματος ἀποκλειόμενα. οὐ μὴν ὀρθῶς γε[2] λέγουσιν, ἀλλ᾽ ἑκάτερα διαιρετέον ὥσπερ εἴρηται.
χρὴ δὲ καὶ πρὸς τὸν τόπον ὁρῶντα ποιεῖσθαι τὰς
ἀποστάσεις· ἐν γὰρ τοῖς ὀρεινοῖς ἐλάττους ἢ ἐν
τοῖς πεδεινοῖς, ἐπ᾽ ἔλαττον γὰρ αἵ τε ῥίζαι προ-

---

[1] U : ξηροί τε Schneider.
[2] aP : τε U N.

---

[1] Those that dislike shade and have long roots.
[2] Cf. CP 2 7. 3–4.

much with regard to the region as with regard to the trees themselves that are being planted, by whether they like or do not like shade and have long or short roots, since the one set[1] must evidently be spaced thin, the others close. Those trees love shade that have very dry fruit with plenty of stone (like pomegranate and myrtle), and that are naturally of open texture, dry,[2] and not long-rooted (like bay).[3] For in the former the fruit becomes astringent when exposed to the sun, and the latter require a screen (as it were) against both cold and hot weather because of their weakness, since they do not interfere with one another with any tangled mass of roots, because the roots are short. And yet some experts tell us to plant all trees wide apart to prevent the roots from entangling together and the trees from getting stunted by getting too little food and being shut off from the wind. They are nevertheless mistaken, and we must distinguish two groups as we have done.[4] But we must also consider the region in calculating the spacing, since in mountainous country the intervals are smaller than on the plain,[5] because on the mountains the

7.2

---

[3] The agriculturists recommended close planting (no more than nine feet apart) for pomegranate, myrtle and bay (*HP* 2 5. 6).   [4] *CP* 3 7. 1.

[5] *Cf. HP* 2 5. 6 (the experts recommend): "that we should calculate the spacing in relation to the country, for in mountain country the distance is less than on the plain."

THEOPHRASTUS

αὔξονται καὶ ἡ κόμη.

7.3 ἅπαν δὲ φυτόν, ὅταν ἐκβλάστῃ,[1] τὸ πρῶτον ἐᾶν δεῖ[2] ῥιζωθῆναι, μηδὲν κινοῦντα τῶν ἄνω (καθάπερ ἐπὶ τῶν ἀμπέλων ποιοῦσιν, ἀφιέντες[3] τὰς ῥάχους), εἶθ' ὅταν ἰσχωσιν,[4] τότε περιαιρεῖν[5] τὰ ἄνω, καταλιπόντα τὰ κάλλιστα καὶ τὰ ἐπιτηδειότατα πεφυκότα (κατὰ γὰρ τὰ δένδρα τοῦτ' ἀναγκαῖον, οὐχ[6] ὥσπερ ἐπὶ τῶν ἄλλων ἀφαιρεῖν ἐνδέχεται πάντα καὶ τέμνειν ἰσόγεων[7])· ἀρριζώτου[8] γὰρ ὄντος ἐὰν περιαιρῇ[9] καὶ κινῇ τις, ἀσθενὲς ὄν, συναισθήσεται[10] μᾶλλον· ἐὰν δ' ἐρριζωμένου καὶ ἰσχύοντος, αὐτό τε ἀπαθὲς ἔσται καὶ τὴν τροφὴν ἀποδώσει πλείονα τοῖς καταλοίποις.

7.4 ἡ δ' ἀναγωγὴ καὶ ἣν καλοῦσιν οἱ πολλοὶ τῶν φυτειῶν[11] παιδείαν οἷον σχηματισμός ἐστι καὶ μόρφωσις τῶν δένδρων ὕψει τε καὶ ταπεινότητι,

---

[1] Wimmer : ἐκβλαστῇ U.
[2] ego : δὲ U^{ar} : U^r N aP omit.   [3] Heinsius : -ας U.
[4] ἰσχωσιν U : ἰσχύωσι Itali : *tenuerit: convalueritque* Gaza.
[5] Schneider : περιαίρειν U.   [6] οὐχ' U : οὐδ' Schneider.
[7] Schneider : ἴσογεων U.   [8] Wimmer : ἀρρίζου τε U.
[9] Schneider : περιαίρηι U.
[10] συνεσθήσεται U : συναιρεθήσεται u : συνθήσεται N : σωθήσεται aP.   [11] U : φυτῶν Gaza, Schneider.

DE CAUSIS PLANTARUM III

roots and foliage do not grow out so far.

The Sapling: Cutting Back

One should allow every slip, when it has come 7.3
out, first to strike root, and not disturb the parts
above ground (so growers do with the vine when
they let the rough shoots grow). Only later, when
the slips have got their roots, should one cut back
the parts above ground, leaving however the parts
that are finest and that grow in the most convenient
positions (since in trees we are forced to leave some
parts behind, because it is not possible as with other
plants[1] to remove the whole upper growth and cut
them down to the ground). For if one cuts back the
parts and interferes with the slip when it has not
yet got rooted, it will, in this weak state, be more
sensitive to the procedure; but if one cuts when it is
rooted and strong, it will suffer no harm itself and
pass more food to the parts allowed to remain.

The Sapling: Training

The training and what is commonly termed the 7.4
schooling in habit[2] is a forming (one might say) and
shaping of trees in vertical and lateral extent and in

---

[1] *Cf. CP* 2 15. 6 with note *a*.
[2] "Habit" renders *phyteia*; for this use of the word *cf. HP* 3 8. 4.

καὶ πλάτει, καὶ τοῖς ἄλλοις. ὡς δ' ἐπίπαν φυτεύουσι[1] τὰ ὑψηλά (πλὴν ὅσα φύσει τοιαῦτα καὶ ἔστι καὶ βούλεται, καθάπερ φοῖνιξ, πεύκη, κυπάριττος)· ἀκαρπότερα γὰρ γίνεται, διὰ πολλοῦ τῆς τροφῆς ἰούσης καὶ ἐνταῦθα καταναλισκομένης. ἀλλὰ δεῖ[2] πολύκλαδα καὶ πολυβλαστῆ ποιεῖν, ἅμα γὰρ τῇ πολυκαρπίᾳ καὶ εὐτρυγητότερα καὶ τὸ ὅλον εὐθεραπευτότερα γίνεται· διὸ καὶ περιαιρετέον καὶ κωλυτέον τὰς μὴ κατὰ καιρὸν βλαστήσεις.

7.5 σχεδὸν δ' ἡ τοιαύτη θεραπεία καὶ κατάστασις ὁμοία τῇ διακαθάρσει τυγχάνει τελείων ὄντων τῶν δένδρων· ὑπὲρ ἧς καὶ δεικτέον πρῶτον τῶν κατὰ τὰς θεραπείας, εἶθ' οὕτω περὶ τῶν ἄλλων, ὥσπερ γὰρ ἐφεξῆς ταῦτα τῶν περὶ τὴν φυτουργίαν ἐστί.

κοινοτάτη μὲν οὖν καὶ μάλιστα ἔνδηλος ἐπὶ τῶν ἀμπέλων τυγχάνει· πάντες γὰρ χρῶνται καὶ πάντες ἀμπελουργοῦσιν, ὥσπερ ὑπ' αὐτῆς ἀναγκαζόμενοι τῆς φύσεως, διὰ τὴν εὐβλάστειαν καὶ πολυβλαστίαν· ὑπὲρ ἧς ἐπειδὴ καὶ ἐν τῇ κλάσει καὶ ἐν τοῖς ἄλλοις διηκρίβωται μάλισθ' ἡ θερα-

---

[1] U: παιδεύουσι Dalecampius: ἀποτέμνουσι Schneider: ἐπιτέμνουσι Wimmer. [2] u aP: δὴ U N.

other ways. On the whole it is the tall trees that are trained (with the exception of those that are naturally tall and strive for height, as the date-palm, pine and cypress), since unless they are shortened they tend to bear less fruit, because the food has a long distance to go and is used up on the way. Instead one should make them many-branched and many-shooted, for besides bearing more they will then become easier to harvest and in general easier to tend, which is why we should also remove and prevent any shoots that grow in inconvenient positions.

The Mature Tree: Pruning

This sort of tending and ordering is similar (one might say) to the pruning of the adult trees. Of the procedures involved in the care of trees we must first discuss pruning and then proceed to the rest, since this subject comes next in order (as it were) to the procedures dealing with the immature tree. 7.5

Pruning, then, is most universally applied to the vine and most clearly seen in its case, since all resort to it and all dress the vine, as if they were compelled to do so by its very nature, owing to the rapid and abundant sprouting. Since the dressing of the vine has been worked out in the greatest detail both in the matter of pruning and in all its other aspects, we must endeavour to study these pro-

THEOPHRASTUS

πεία, πειρατέον αὐτὰ καθ' αὑτὰ θεωρεῖν ὕστερον·
νῦν δ' ὑπὲρ τῶν ἄλλων λέγομεν.[1]

7.6 οὐδὲ γὰρ οὐδ' ἐπὶ τούτων μικρόν ἐστι τὸ κατα-
στήσασθαί πως τὰ δένδρα καὶ κακῶς πεφυκότα
καὶ τὰ[2] ἐμποδίζοντα[3] τοὺς καρποὺς ἀφαιρεῖν
(ἐμποδίζει γὰρ οὐ ταῦτα μόνον, ἀλλὰ καὶ τὰ
ἀπηρτημένα καὶ αὖα καὶ[4] ὅλως τὰ παρακαίρως
πεφυκότα)· διὸ δεῖ ταῦτ' ἐξαιροῦντα μετατιθέναι
τὴν βλάστησιν εἰς τὸ δέον. ἔοικεν γὰρ ὥσπερ[5]
ὀχετεία τις εἶναι τῆς[6] τροφῆς τῶν δένδρων, ὅπως
ἂν ἄγῃ[7] τις· εἰς γὰρ τὰ καταλειπόμενα ῥεῖ καὶ
ταῦτ' αὔξει.

7.7 τοῦτο δὲ ξυμβαίνει διότι τοῖς μέρεσιν ἄριστα
πέφυκεν, καὶ ἔτι προσεπιβλαστάνει κατ' ἐνι-
αυτόν· ἐπεὶ καὶ εἴ τις ἀεὶ βούλεται κολούειν[8] τὰ
ἄνω, πρὸς τὰς ῥίζας ἡ τροφὴ πᾶσα φέρεται,
κἀκεῖναι λαμβάνουσιν αὔξησιν ὥστε συνταρροῦσ-
θαι[9] τὰ χωρία, καὶ τέλος ἀναυξῆ γίνεται τὰ ἄνω,
πάσης ἐνταῦθ' ὡρμηκυίας τῆς τροφῆς, οἷον γὰρ
ἤδη φύσις γίνεται χρονισθέντων.

[1] U N P : λέγομεν u Gaza a.   [2] [τα] Schneider.
[3] aP : ἐμπαίζοντα U N.   [4] αὖα καὶ Schneider : αὐτὰ U.
[5] Gaza a : ὥστε περ U N P.   [6] [τῆς] Schneider.
[7] Gaza, Schneider : ἀνάγκη U.
[8] U^c : κωλύειν U^{ac}.   [9] Gaza (*radicibus impleantur*),
Scaliger : συνταραττεσθαι U.

DE CAUSIS PLANTARUM III

cedures later in a separate section[1]; at present we deal with the other trees.

So too in the grown tree: it is of no small importance to reduce the tree to a certain order and remove the parts that are ill grown and that impede the fruit (for not only do the ill-grown parts impede it but so too do the parts that grow away from the rest and the deadwood and indeed all parts that have grown in an inopportune way). We must therefore remove these and thus shift the sprouting to where it is wanted, since this removal appears to be a channelling (one might say) of the tree's food by guiding it as one pleases, the food flowing into the parts allowed to remain and making them grow.

7.6

This happens because of all plants trees are the best endowed in their parts, and moreover have a second growth of shoots every year.[2] Indeed if one wishes to cut back all the new upper growth as it grows out, all the food moves to the roots, and these grow to such an extent that the orchards are matted with them, and finally the parts above cease to grow, all the food having taken its course to the parts below, for when training has lasted long enough it turns as it were into nature.[3]

7.7

---

[1] *CP* 3 11. 1–3 16. 4.

[2] *Cf. HP* 3 5. 4: "... but the second sproutings in the dog-days and at the rising of Arcturus are (one may say) common to all (*sc.* trees)..."

[3] Habit becomes nature: *cf. CP* 2 5. 5 with note.

THEOPHRASTUS

ἡ μὲν οὖν ἄμπελος ἀεὶ τὴν τομὴν ἐπιζητεῖ κατ' ἐνιαυτόν, διὰ τὸ εὐαξές· τὰ δὲ ἄλλα, τὰ μὲν παρ' ἔτος, τὰ δὲ διὰ τετραετίας,[1] οὐδὲν γὰρ οὕτως εὐαξές· ἐπεὶ εἴ γέ τι τοιοῦτον εἴη, δέοιτ'

7.8 ἄν. <ὧν> ἀφαιρεῖν[2] ἀπὸ τῶν ὑγρῶν περίεργον, ἅμα δ' ἡ ἕλκωσις πόνον παρέχει καὶ κακοῖ τὰ δένδρα, διὸ καὶ μετὰ τὴν διακάθαρσιν εὐθὺς οἴονται[3] δεῖν κοπρίζειν καὶ τὴν ἄλλην ἀποδιδόναι θεραπείαν, ὅπως ἐπαναλάβωσι τῇ τροφῇ τὴν κακοπάθειαν· ἀλλὰ μόνον τό γε συνεχὲς ἐν τῇ τῶν αὔων ἀφαιρέσει ποιεῖσθαι· ταῦτα γὰρ οὔτε πόνον ποιεῖ τῇ ἑλκώσει, κωλύει τε προσηρτημένα τὰς τροφάς.

7.9 ἐν δὲ τῇ διακαθάρσει τά τε μὴ κάρπιμα διαιρετέον, καὶ ὅσα τῶν ἑτέρων αὔξησιν ἀφαιρεῖται, καὶ ὅσα διαπέφυκεν (ἐν τοῖς ἔξω γὰρ δεῖ τὴν βλάστησιν εἶναι), καὶ ἔτι τὰ πυκνὰ καὶ ἀλλήλοις ἐπιβάλλοντα, καὶ ὅσα τὴν ἔκφυσιν ἐκ τῶν μέσων

---

[1] u a (in an omission in P) : τετραἔτειας U : τετραετείας N.
[2] <ὧν> ἀφαιρεῖν ego : ἀφαιρεῖν <δ' ἀεὶ> Schneider.
[3] u : οἷον τε U.

---

[1] As opposed to removing deadwood.
[2] *Cf. HP* 2 7. 2: "All trees need pruning, since they are better when the deadwood, which is (as it were) something

DE CAUSIS PLANTARUM III

Whereas the vine requires the pruning to be annual because its growth is so rapid, the rest require pruning in some cases every other year, in others every third year, none being so rapid a grower as the vine; indeed if any should grow so rapidly, it would need pruning annually. But to keep pruning from the fresh wood[1] in these others is needless trouble, and then too the wounding is an affliction and a hardship, which is why the experts believe that immediately after pruning one should manure the tree and give it the other attentions, to let it make up for the hardship by good feeding, and that only the removal of deadwood should be maintained without interruption, since deadwood can be removed without inflicting wounds and when it is left on the tree prevents feeding.[2]

7.8

In pruning the branches of the grown tree we must distinguish from the rest[3] the parts that do not bear, those that prevent the growth of other parts, and those that grow in between[4] (since the sprouting should be at the periphery);[5] further we must distinguish the parts that are too crowded and cross one another, and those that grow from the

7.9

foreign, is removed, and it furthermore obstructs growth and feeding."

[3] And remove.

[4] That is, they grow in between other branches and stop short of the periphery.

[5] Because the fruit will then be exposed to sun and air.

THEOPHRASTUS

ἔχει, πάντα γὰρ ταῦτα καὶ τὸ πνεῦμα καὶ τὸν ἥλιον ἀφαιρεῖται· δεῖ δὲ καὶ εὔπνουν εἶναι καὶ πρόσειλον τὸ δένδρον, διὸ καὶ οὐ κακῶς οἱ οὕτως ῥυθμίζοντες ὥστε πρὸς μεσημβρίαν βλέπειν (καθάπερ οἱ τὰς συκᾶς[1] καὶ τὰ ἄλλα καὶ μάλιστα τὴν ἐλάαν).

7.10   ἐπεὶ δ' ἐπίπονος ἡ διακάθαρσις διὰ τὰς πληγάς, διὰ τοῦτ' οὐ τὴν τυχοῦσαν ὥραν, ἀλλὰ ⟨τὴν⟩[2] μετὰ Πλειάδα ληπτέον· ἰσχυρότατα γὰρ τότε καὶ μάλιστα αὐτῶν[3] τὰ δένδρα, καταναλωκότα τὸ ὑγρὸν εἰς τοὺς καρπούς, ἕτερον δὲ οὐδέπω δεδεγμένα·[4] τὰ δ' ὀψικαρπότερα δῆλον ὅτι κατὰ λόγον, μετὰ τὴν τῶν καρπῶν ἀφαίρεσιν. ἴδιον δ' ἐπὶ τῆς συκῆς· μόνη γὰρ διακαθαίρεται μικρὸν πρὸ τῆς βλαστήσεως. αἴτιον δ' ὅτι τότε μάλιστα εὐσύμφυτον·[5] δεῖ δὲ τοῦτο[6] σπεύδειν, ἀσθενὲς γὰρ ὂν καὶ μανόν, πονεῖ μάλιστα διὰ τοῦ μετοπώρου καὶ περὶ Πλειάδα, ξηροτέρας οὔσης τῆς ὥρας οὐ δυναμένη[7] συμφῦναι, σήπεταί τε παραρρέοντος τοῦ ὕδατος, καὶ ὑπὸ τῶν χειμώνων κακοπαθεῖ, καὶ

[1] Gaza, Itali : οἰκείας U : συκέας u : συκίας N : σικύας aP.
[2] Schneider.
[3] Scaliger (cf. ἑαυτοῦ CP 3 4. 4) : αυτῶν U.
[4] Gaza (susceperint), Heinsius : δεδειγμένα U.
[5] U N P : -ος a.   [6] Gaza (quod), Scaliger : ταυτὸ U.

DE CAUSIS PLANTARUM III

middle of the branch, for all shut out wind and sun, and the tree must be both well-aired and exposed to the sunshine. This is why it is a good precept to discipline it so that it faces south, as is done with the fig and the rest,[1] especially the olive.

Since branch pruning is a hardship because of 7.10 the wounds we must choose for it not any season that we please but the season after the setting of the Pleiades, since the trees are then strongest and most their own selves,[2] having expended their fluid on the fruit and not yet received a new supply. Later-fruiting trees must be pruned at a correspondingly later time, after the fruit has been harvested. The fig-tree is a special case: it is the only tree that is pruned shortly before it sprouts. The reason is that the wounds are then most easily closed, and this must be our aim, since the tree is a weak and open-textured one and suffers more than any other in the course of the autumn and at the setting of the Pleiades, since the fig is unable to close its wounds when the season is drier, and when the rains come it decomposes when the water gets into the cuts, and

---

[1] That is, the rest mentioned in this connection by the agriculturists; it is equivalent to "etc."
[2] For the phrase *cf. CP* 3 4. 4; 4 3. 4. It indicates that the tree is not responding to some seasonal urge or to some emergency.

---

[7] U^r : δυναμένης U^{ar} N aP.

τὸ ὅλον διαφθείρεται.

τοῦτο μὲν οὖν ἴδιον τῶν ἀσθενῶν καὶ μανῶν.

7.11 οὐκ ἴση δὲ πάντων ἡ ἀφαίρεσις, ἀλλὰ πλείων[1] τῶν μᾶλλον δεομένων· δεῖται[2] δὲ τὰ εὐβλαστοῦντα πολλῆς γενομένης,[3] οἷον ἐλάα δοκεῖ καὶ ῥόα καὶ μύρρινος· ὅσῳ γὰρ ἄν τις[4] ἐλάττω τούτοις καταλίπῃ, βέλτιον βλαστάνει καὶ τοὺς καρποὺς φέρει βέλτιον· τοῦτο δ' ὅτι πυκνόβλαστα [ἐστὶν][5] καὶ λεπτόβλαστα [ἐστιν][6] καὶ ταχὺ παραυαινόμενα (διὸ καὶ φρυγανικώτατα πάντων τῇ προσόψει).

7.12 χρὴ δὲ καὶ[7] ὅσα πρὸς τὰς τομὰς ἀσθενῆ, καθάπερ ἄπιος καὶ μηλέα καὶ εἴ τι ἄλλο ξηρὸν καὶ λεπτόφλοιον,[8] καὶ γὰρ ταῦτα πονεῖ· διὸ τὰ τοιαῦτα[9] τούτων ἢ ταῖς χερσὶν ἀφαιρεῖν, ὥσπερ ἐλέχθη τε καὶ[10] κελεύουσιν, ἢ τοῖς σιδήροις[11] ὡς ἐλαφρότατα· κίνδυνος γὰρ ἅμα τῷ πόνῳ διὰ τὴν ἕλκωσιν. ὥσπερ γὰρ καὶ τῶν ῥιζῶν ἐν τῇ σκαπάνῃ τιτρωσκομένων χεῖρω γίνεται, πολλάκις δὲ καὶ

---

[1] U^{ac} u : πλεῖον U^c. [2] Schneider : δεῖ U.
[3] U : γινομένης Wimmer.
[4] aP : τι U. [5] aP.
[6] Schneider. [7] δὲ καὶ U : δ' ἐλάττονος Wimmer.
[8] λεπτόφλοιον < . . . > Schneider. [9] U : αὖα Schneider.
[10] [τε καὶ] Schneider : τινὲς Wimmer.
[11] u : σιδηροῖς U.

it suffers from the cold weather and is even killed.

So this is a special case applying to trees that are weak and of open texture.

The amount removed is not the same for all trees, 7.11 but greater for those that require it more, and rapid sprouters require that the removal should have been extensive, as is held to be the case with the olive, pomegranate and myrtle; for the fewer parts one has left in these the better the tree sprouts and bears.[1] The reason for this is that the trees have crowded shoots that are thin and quickly wither (which is why of all trees they bear the closest resemblance to undershrubs).

We must also prune trees that are too weak to 7.12 take cutting well, as pear, apple and other dry trees with thin bark (for these too[2] suffer). So one must remove such parts from them either with the bare hands (as we said[3] and as experts advise) or, if we use iron tools, we must cut as gently as possible, since not only is hardship involved, but danger too from the wounds. For just as trees deteriorate and often even get diseases and wither away when the

---

[1] *Cf. HP* 2 7. 2: "Androtion says that the trees requiring the greatest amount of pruning are the myrtle and olive, for the fewer parts you leave, the better these will sprout and the more abundant will be the fruit they bear"; *HP* 2 7. 3: "Androtion says ... that olive, myrtle and pomegranate ... require the most pruning ..."

[2] Like the fig: *cf. CP* 3 7. 10.   [3] *CP* 3 2. 2.

65

THEOPHRASTUS

νοσεῖ καὶ ἀφαυαίνεται, τὸν αὐτὸν τρόπον οἴεσθα χρὴ καὶ ἀπὸ τῶν ἄνωθεν ἐν τοῖς μὴ δυναμένοι φέρειν.

καὶ περὶ μὲν καθάρσεως ἀρκείτω τὰ εἰρημένα.

8.1 ῥιζοτομεῖται δὲ μετὰ τὴν φυτείαν καὶ ὅταν νέα πάνθ' ὡς εἰπεῖν, ὅπως τε κατὰ βάθους ὠθῶν ται καὶ πλείω λαμβάνωσιν αὔξησιν, μάλιστα δ ὧν[1] ἐπιπολῆς αἱ ῥίζαι, καθάπερ ἐλάα καὶ ἄμπε λος· ὅταν δὲ πρεσβύτερα γένηται, τὰς μὲν ἐπε τείους καὶ ὅλως τὰς ἐπιπολῆς ἀφαιρετέον, ὅπω αἱ κάτω πλείους καὶ ἰσχυρότεραι γίγνωνται, φθεί ρουσι γὰρ ἐκείνας αὗται καὶ ἐμποδίζουσιν, αὐταὶ δ' οὐ δύνανται παρέχειν τροφήν, ἀλλὰ καὶ πονοῦσ καὶ ὑπὸ τοῦ ἡλίου καὶ ὑπὸ τοῦ ψύχους· ἀφαιρετέο δὲ καὶ τὰς αὔας, προσηρτημέναι γὰρ καὶ ἄλλως λυμαίνονται καὶ σκώληκας ἐμποιοῦσι τοῖς δέν δροις.

8.2 τὰς δὲ ἄλλας οὐ κινητέον· οὐδὲ γὰρ τὴν δια κάθαρσίν ἔστιν ὥσπερ τῶν ἄνω ποιεῖσθαι, καὶ γὰρ τὸ γυμνοῦν ἐπὶ πλέον χαλεπόν, καὶ ὅλως οὐκ εὔ σημον ὅθεν ἀφαιρετέον· ἅμα δὲ καὶ ὁποθενοῦν

[1] u : ὡς U.
[2] Wimmer : αὐτοῖς U.

DE CAUSIS PLANTARUM III

roots are wounded in spading, in the same way we must suppose that ill effects arise from what is done to the parts above ground in trees that are unable to bear it.

Let this discussion suffice for pruning.

The Mature Tree: Root-Pruning

After planting and when they are young practically all trees are root-pruned, to make the roots thrust deep and grow larger, especially trees with shallow roots, like the olive and vine. When they get older we should remove the roots produced during the year and indeed all roots at the surface, so that the lower roots may increase in number and vigour, since the surface roots destroy and interfere with these, but cannot supply food themselves, and suffer instead from the sun and the cold. We must also remove the withered roots, since when they remain they hurt the trees in various other ways and give them worms.

But the other roots should not be disturbed. Indeed one cannot prune them as one does the parts above, for to lay them bare to any extent is difficult; and in any case it is not easy to tell at what point to prune. Then too, no matter from what point a root

8.1

8.2

---

[3] Schneider : πόθεν οὖν U.

φυομένη, μόνον δὲ ἰσχύουσα, τὴν τροφὴν ἀποδώσει, κάλλιον δ' ἐὰν πανταχόθεν· τὸ δὲ δὴ μῆκος καὶ πλάτος, ὅσῳ[1] ἂν ᾖ πλέον, ὠφελιμώτερον, ὥστε τὰ πρὸς τὴν βλάστησιν καὶ τὴν εὐτροφίαν μόνον δεῖ παρασκευάζειν.

8.3 περὶ δὲ ἀρδεύσεως καὶ ὑδάτων σχεδὸν εἴρηται πρότερον, ὅσα γὰρ ἐν τοῖς οὐρανίοις ὠφέλιμα κατὰ ποιὰς[2] ὥρας ἢ[3] ἔτους, ἢ νύκτωρ ἢ μεθ' ἡμέραν, δῆλον ὅτι καὶ ἐν τοῖς ἐπιρρύτοις καὶ ὀχετευομένοις ὁμοίως ἔσται. τοῦ δὲ θέρους εὐλόγως μάλιστα χρῶνται, διὰ τὴν σπάνιν τῶν ἐκ Διός, καὶ ἅμα πρὸς τὴν ἐκτροφὴν τῶν καρπῶν ἀναγκαῖον· οἱ[4] δὲ καὶ οἴονται τότε μάλιστα δεῖσθαι, πλείστης ἀφαιρέσεως καὶ κυριωτάτης γινομένης.

8.4 ἰσχυρὸν δ', ὥσπερ ἐν τοῖς ἄλλοις, καὶ ἐν τούτοις τὸ ἔθος, οἷον γὰρ φύσις γίνεται· διὸ καὶ τὰ ἐν τοῖς ξηροῖς καὶ ἀνύδροις οὐθὲν ζητεῖ πλὴν τὸ ἀναγκαῖον, καὶ χείρω γίνεται βρεχόμενα, καθάπερ τὰ

---

[1] Schneider : ὅσων U : ὅσον u.
[2] u : ποίας U N aP.
[3] U (season *either* of the year or of the day) : τοῦ Schneider.
[4] aP : εἰ U N.

comes, so long as it is vigorous it will supply its quota of food, and the roots will do this better if they grow from all sides of the tree. As for length and lateral extent, the greater the better. All we need to do then is to see that these roots come out and are well fed.

Watering

What is to be said about watering and the kinds of water has to all intents been said before,[1] since all the benefits from rain that come from its falling at one season or the other of the year or in the dark or light hours of the day will, it is evident, be equally present in water from streams and irrigation ditches. It is reasonable that growers resort to watering chiefly in summer because of the scarcity of rain at the very time when water is indispensable for bringing the fruit to maturity. Some authorities moreover believe that watering is then most needed by the trees because they are undergoing the greatest and most important loss of their fluid.

8.3

Here as elsewhere the force of habit is great, for it turns (as it were) into nature.[2] This is why trees in dry and waterless country do not seek water beyond the indispensable minimum and deteriorate on being watered, just as trees do that are accus-

8.4

[1] *CP* 2 2. 1–4.
[2] *Cf. CP* 2 5. 5 with note *c*; *CP* 3 3. 7.

THEOPHRASTUS

εἰωθότα μὴ[1] βρέχεσθαι.[2] καὶ κίνδυνος δ' ἐὰν[3] μὴ συνεχῶς ἀποδιδῷ, βρέχειν[4] ἀρξάμενος·[5] ἐκείνην τε γὰρ ἀφαιρεῖται τὴν ἰσχὺν καὶ τὴν τροφήν, καὶ ἑτέραν οὐ διδούς, ἀμφοτέρων πλείω διαμαρτάνει, ὥστε πολλάκις ὑπὸ τῶν καυμάτων αὐαίνεται. σχεδὸν δὲ οἱ καρποὶ χείρους τῶν γε πλεῖστον[6] δεχομένων· ἐὰν δὲ συμμέτρως, ἅμα[7] πλείονες[8] καὶ καλοὶ[9] μᾶλλον.[10]

9.1   περὶ δὲ κοπρίσεως ἀκολούθως δῆλον ὅτι τῇ διακαθάρσει[11] κατὰ[12] τοὺς χρόνους, ἄλλως τε εἰ καὶ μετ' ἐκείνην εὐθὺς κοπρίζειν δεῖ· ἅμα δὲ καὶ οὐδ' ἂν δύναιντο δέχεσθαι συνεχῶς, ἀλλ' ἀφαυαίνονται διὰ τὴν θερμότητα. τὸ γὰρ τὴν ψιλὴν κοπρίζειν οὐχ ὅμοιον· ἐν ἐκείνῃ μὲν γὰρ εἰς πολλὰ μερίζεται καὶ ἐξαναλίσκεται τὰ δεόμενα[13] τῶν σπερμάτων, ἐνταῦθα δὲ εἰς αὐτὰς τὰς ῥίζας δύεται, καὶ οὐ

---

[1] U^c : U^t omits.   [2] U^{cc} from βρεχεσεσθαι (?).
[3] N : δὲ ἂν U^c (δὲ from δ'ε) aP.
[4] after βρέχειν two letters erased U^c.
[5] u^{ac} (αρξαμενος U) : ἀρξαμένοις u^c.
[6] u : πλείστων U N aP.   [7] ego : αλλα U.
[8] u : πλείονι U : πλεῖον N P : καὶ πλεῖον a.
[9] ego : πολυ U : πολλοὶ u : πολλῶ N aP.
[10] U : καλλίους Wimmer.
[11] Schneider : τῇ ἰδίᾳ καθάρσει U.

tomed to not getting watered. And it is actually risky to start watering and not keep it up, since you deprive them of their earlier strength and way of feeding and then by withholding any other make a greater mistake than either the steady waterer or steady non-waterer, so that the hot weather often withers the trees. One may say that in the trees, those at least that get the greatest amount of water, the fruit deteriorates[1]; whereas if it is received in the right amount the fruit is more abundant and tends to be finer.

Manuring

As for manuring, it should correspond to pruning in the time of application,[2] especially as we are told to manure immediately after pruning.[3] Then too trees could not stand constant manuring, but dry out from the heat. For manuring field crops is not a similar case: in the field the manure is divided up into many portions and used up by the grain seeds that need it, whereas with trees it sinks to the roots,

9.1

---

[1] *Cf. HP* 2 7. 1: "When watered the fig sprouts faster but its fruit deteriorates ..."
[2] *Cf. CP* 3 7. 10.
[3] *Cf. CP* 3 7. 8.

---

12 ego (λεκτέον Schneider) : μετα U.
13 ego (διαδιδομένη Schneider : τὰ δεχόμενα Wimmer) : τὰ δὲ οὐ μόνον U.

THEOPHRASTUS

δύναται πᾶν διαδιδόναι τοῖς καρποῖς, οὐχ ὁμοίως εὔτροφον <ὂν>[1] οὐδὲ πολύχουν, ὥστε πάλιν ἄλλης καὶ ἄλλης ἐπιγινομένης, ὡς ἐκθερμαινόμενον, αὐαίνεται. ἐπεὶ[2] καὶ ὁ σῖτος, ἄν τις κατακόρως χρῆται καὶ <μὴ>[3] ἐπιγίνηται πλῆθος ὕδατος· διὸ καὶ ταῖς ἐπομβρίοις[4] χρησιμώτερος ὅ γε συνεχὴς καὶ πλείων κοπρισμός, ἐν δὲ ταῖς αὐχμώδεσι καὶ λεπταῖς ὁ σύμμετρος. διὰ ταῦτα δὲ ἴσως οὐδὲ ἡ δριμυτάτη τοῖς δένδροις ἁρμόττει. καίτοι δόξειεν ἂν ἄτοπον εἶναι τὸ μὴ τοῖς <ἰσχυροτάτοις τὴν>[5] ἰσχυροτάτην· ἀλλ' ἐν τῷ προειρημένῳ τὸ αἴτιον ληπτέον. ἐπεὶ[6] καὶ τοῖς λαχάνοις διὰ τοῦθ'[7] ἁρμόττει, <καὶ>[8] διὰ τοῦτο[9] μεθ' ὑδρείας ἡ χύλωσις καὶ ὅλως, ὅτι πολλῷ τῷ ὑγρῷ χρῶνται καὶ καθ' ἡμέραν, καὶ αὐτὰ φύσει ὑγρά· φυλλοφορεῖν γὰρ ἐθέλουσιν, οὐ καρπογονεῖν αὐτά.

οὐ μὴν ἀλλ' ἐνίαις γε[10] χρῶνται καὶ πρὸς τὰ δένδρα τῶν ἰσχυροτέρων, καὶ μάλιστα εἰς μαλακό-

---

[1] Wimmer. [2] u : ἐπι U.
[3] U : φυτῶν Gaza, Schneider.
[4] ego (ἐπόμβροις Schneider) : -ίαις U.
[5] Gaza, Schneider.
[6] U^c from ἐπί.
[7] [δια τούθ'] Heinsius. [8] ego.
[9] [δια τουτο] Schneider. [10] Schneider : τὲ U.

and not every tree has the power to apportion it among the fruiting parts, since a tree is not so good a feeder or so prolific a producer as grain. So with more and more manure accumulating, the tree becomes over-heated and withers away. Indeed cereals wither too if manuring is carried to an extreme and not followed by heavy rain; and this is why sustained and copious manuring is more useful in rainy country, but moderate manuring in dry country and thin soil. This is perhaps also the reason why the most pungent manure does not agree with trees.[1] Yet it might be thought strange that the strongest manure should not agree with the strongest of plants. But the explanation is to be taken from what we said before.[2] Indeed the strongest agrees with vegetables for this reason (and this is why there is such a thing as the use with them of liquid manure, and so manuring and watering them at once): gardeners use plenty of water and use it daily, and the plants are themselves by nature fluid (since cultivators want leaves from them, not fruit).[3]

Nevertheless some of the stronger manures are also used with trees, chiefly to obtain softer stones

---

[1] *Cf. CP* 3 6. 2.
[2] *CP* 3 9. 1.
[3] *Cf. HP* 7 5. 1: "All vegetables except rue like water and manure ... Gardeners ... also use fresh human dung in making liquid manure."

THEOPHRASTUS

τητα καὶ μεταβολὴν τῶν καρπῶν, οἷον ὑείᾳ[1] πρὸς τὸ[2] γλυκαίνειν καὶ ἀπυρήνους[3] ποιεῖν τὰς ῥόας, καὶ τὰς ἀμυγδαλᾶς ἐκ πικρῶν γλυκείας, καὶ ὡς παρὰ τοὺς μυρρίνους κελεύουσιν ἰσχυροτέραν ἔτι παραβάλλειν, οἷον τὴν βυρσοδεψικήν, καὶ οὖρον παραχεῖν[4] ὅταν ἐκβλαστήσωσιν,[5] ὡς ἂν ἀπυρήνων γινομένων· οἱ δὲ καὶ τῇ ἐλάᾳ συμφέρειν οἴονται[6] πρὸς εὐκαρπίαν.

9.4 ἔοικεν δὲ τὸ ἐν τῇ ἀρχῇ μέγα διαφέρειν, εἴ γε καὶ τὰ τῶν σικύων σπέρματα γάλακτι βρεχόμενα καὶ μελικράτῳ γλυκυτέρους ποιεῖ. νομίσειε δ' ἄν τις ταῦτα καὶ ἐπὶ τῶν ἄλλων γίνεσθαι λαχάνων, ἀλλ' ἴσως τὸ δριμὺ χρησιμώτερον. ἡ δὲ ῥίζα τῶν δένδρων ἀρχή τις οὖσα, καὶ τὰ περικάρπια συνεξομοιοῖ τὰ ἀφ' ἑαυτῆς. περὶ μὲν οὖν τούτων καὶ πρότερον εἴρηται.

τὴν δὲ χρησίμην ἀεὶ πρὸς τὰς δυνάμεις ἀποδοτέον, τῆς μὲν δριμυτέρας ἐλάττονα τῆς δὲ μαλακωτέρας πλείονα.

---

[1] Wimmer (*suillo* Gaza : τῇ ὑείᾳ Scaliger) : τῇ U.
[2] N aP (το U) : τῷ u.
[3] Gaza : ἀπύρους U.
[4] U^r : παρασχεῖν U^{ar} N aP.
[5] N aP : ἐμβλαστήσωσιν U.
[6] u : οἷον τε U.

DE CAUSIS PLANTARUM III

in the fruit and change the taste, as swine manure is used to make pomegranates sweet[1] and produce "stoneless" fruit, and to change almonds from bitter to sweet,[2] and with myrtles[3] we are told to use even stronger manure, such as tanner's, and pour urine round them when the shoots come out, since then it appears the fruit will have no stone. Some[4] think that strong manure also helps the olive to bear well.

It appears that what happens at the beginning makes a great difference,[5] inasmuch as when cucumber seeds are soaked in milk[6] or hydromel[7] they make sweeter cucumbers (one would suppose that this would happen with the other vegetables as well, but perhaps pungency is here a more desirable character), and since the root is a kind of beginning of trees, it gives the pericarpia that come from it a quality similar to its own. But this has been discussed earlier.[8]

As for the manure that is useful, one must always apply it in amounts that vary with its differences of potency, less of the more pungent, more of the less.

9.4

---

[1] *Cf. CP* 2 14. 2 with note *a*.   [2] *Cf. CP* 2 14. 2.

[3] In *HP* 2 7. 3 Androtion is quoted as saying that olive, myrtle and pomegranate require the most pungent manure.

[4] No doubt Androtion: see the preceding note.
[5] *Cf. CP* 3 2. 7.   [6] *Cf. CP* 2 14. 3 with note *a*.
[7] *Cf. CP* 5 6. 12 (honey).   [8] *CP* 2 14. 3.

THEOPHRASTUS

9.5 δεῖ δὲ καὶ τὴν πρόσφορον ἑκάστοις τῶν δένδρων μὴ ἀγνοεῖν· οὐ γάρ, ὥσπερ ὕδωρ τὸ αὐτὸ πᾶσιν, οὕτω καὶ κόπρος, ἀλλ' ὥσπερ τὰ ἐδάφη πρὸς ἕκαστον οἰκεῖα, οὕτω καὶ ἡ κόπρος. ἐπεὶ καὶ κατὰ τὰς ἡλικίας τῶν δένδρων ἐστίν τις διαφορά· διὸ καὶ φυτεύοντες ὑποβάλλουσιν εὐθὺς (ὥσπερ ἐλέχθη) τὴν τῶν λοφούρων, ὅτι κουφοτάτη, τὸ δ' ἀσθενὲς κουφοτάτης δεῖται. τὸ δ' ὅλον ἐν ταῖς θεραπείαις τῶν δένδρων ὅσα μὲν κοινὰ πᾶσίν ἐστιν, ταῦτα τῷ ποσῷ καὶ τῷ ποιῷ διοίσει καὶ τοῖς καιροῖς,[1] οἷον τομὴ διακάθαρσις[2] σκαπάνη κόπρισις. ταῖς δ' οὖν ἀμπέλοις ἢ διὰ τεττάρων ἢ πλειόνων ἐτῶν παραβάλλουσι[3] κόπρον, οὐ γὰρ δύνανται φέρειν δι' ἐλαττόνων, οὐδ' ἐστὶ βοήθεια, καθάπερ τοῖς δένδροις, ἡ ὕδρευσις, ἀλλ' ἐκκαίονται. διὸ (καθάπερ ἐλέχθη) ταῖς ἐπομβρίοις[4] χώραις συμφέρει[5] μᾶλλον· ἄλλως[6] γὰρ κίνδυνος, μὴ ἐπιγινομένων τῶν ἐκ Διός.

[1] Schneider : καρποῖς U.
[2] u : διακάθαρσεις U.
[3] U aP^c : περιβάλλουσι u : παρεβάλλουσι N P^{ac} (?).
[4] Heinsius (ταῖς ἐπόμβροις Schneider) : ταῖς ἐπομβρίαις U N aP : τὰς ἐπομβρίας u.
[5] v : -ειν U N aP.
[6] u a : ἄλλος U N P.

DE CAUSIS PLANTARUM III

We must also not be ignorant of what manure is suited to each tree, since it does not hold of manure, as of water, that the same is good for all, but rather, just as soils vary in their appropriateness to each tree, so too does manure.[1] Indeed even the different ages of a tree make a certain difference in the manure to be used; this is why at the very beginning, when the tree is planted, growers put in pack-animal manure (as we said)[2] because it is lightest, and the weak plant requires the lightest. And in general in the care of trees the measures used for all will differ for each in quantity and quality and times, as cutting back, pruning, spading and manuring.[3] With the vine at all events manure is applied every three years or at even greater intervals, since the vine cannot stand more frequent manuring, and watering is here no remedy, as it is with trees, but the vine is burnt and dies out. This is why manuring (as we said)[4] is better for rainy countries; otherwise there is risk if no rain follows.

9.5

---

[1] Referred to at *CP* 5 15. 3. *Cf. HP* 2 7. 4: "Manure does not suit all trees alike nor does the same suit all; for some require it to be pungent, some require it to be less so, and some require it to be quite light." For the variation of soils *cf. CP* 2 3. 4–5.   [2] *CP* 3 6. 1–2.

[3] *Cf. HP* 2 7. 1: "In tillage and tendance some measures are common to all trees, some peculiar to a kind. Common to all are spading, watering and manuring, and further pruning and the removal of deadwood."

[4] *CP* 3 9. 2.

THEOPHRASTUS

10.1 σκαπάνη δὲ πᾶσι συμφέρει, τά τε γὰρ ἐμποδίζοντα καὶ τὰ παραιρούμενα <τὰς>[1] τροφὰς ἐξαιρεῖ καὶ αὐτὴν τὴν γῆν ἐνικμοτέραν ποιεῖ καὶ κουφοτέραν· ἔτι δ' ὁ ἀὴρ ἐγκαταμιγνύμενος (ἀνάγκη γὰρ ἐγκαταμίγνυσθαι κινουμένης[2]) ἰκμάδα τέ τινα δίδωσιν καὶ παρέχει τροφήν. διὸ καὶ τὴν αὐχμώδη καὶ ἄνυδρον σκάπτειν δεῖ[3] καὶ μεταβάλλειν πολλάκις, ὥσπερ καὶ πρότερον εἴρηται· συμφέρει δ' ἡ σκαπάνη καὶ τοῖς ἑλώδεσι καὶ τοῖς ἐφύδροις. καίτοι δόξειεν ἂν ἄτοπον εἰ τοῖς ἐναντίοις· ἀλλ' οὐδὲν ἄτοπον, τὴν μὲν γὰρ ξηραίνει, τὴν δ' ὑγραίνει, δεῖται δὲ ἑκατέρα[4] τοῦ ἐναντίου.

10.2 τροφῆς δὲ πλείονος καὶ βελτίονος γινομένης αὐτὸ τὸ δένδρον εὐθενεῖ καὶ οἱ καρποὶ καλλίους· ποιεῖ γὰρ καὶ τοὺς πυρῆνας ἐλάττους, καὶ ὅσων[5] ξυλώδη καὶ δερματικὰ τὰ ἐκτός (οἷον ἀμυγδαλῆς καρύας Εὐβοϊκῆς), τούτων λεπτότερα ταῦτα, τὰ δ' ἐντὸς[6] μείζω· τὸ γὰρ ὅλον ἡ εὐτροφία διυγραί-

---

[1] u : U omits between 195ʳ and 195ᵛ.
[2] U : κινούμενος u N : κινούμενον aP.   [3] u : δὴ U.
[4] u : ἑκάτερα U.   [5] Gaza, Wimmer : ὅσα U.
[6] Gaza, Itali : ἐν τοῖς U.

---

[1] Cf. HP 2 7. 5: "The authorities hold spading good for all trees, as hoeing for the smaller plants, since they then get better fed."

DE CAUSIS PLANTARUM III

Trees: Spading

Spading is good for all,[1] since it removes the blockers and interceptors of the supply of food and makes the earth itself damper and lighter; again, air gets mixed in with the earth, as it must when the earth is turned up, and imparts a certain moisture and so provides food. This is why one must spade even dry and waterless ground and turn it up frequently (as we said before)[2]; but spading is also good for land that is marshy and has surface water. Yet it might seem strange that it is good for opposite kinds of land. But there is nothing strange here, since it dries the wet ground and wets the dry, and each requires its opposite.

10.1

With more and better food the tree enjoys well-being itself and its fruit is finer, for such food reduces the size of the stones and where the outside is woody or leathery (as in almond and sweet chestnut) makes this thinner and the inside larger. For good feeding in general soaks with fluid and makes

10.2

---

[2] *HP* 2 7. 5: "Dust too is considered to be nutritious to certain plants and to make them flourish, as it does to the grape-cluster, and growers therefore dust it frequently; some also hoe the fig-trees where this is needed. At Megara they also hoe and cover with dust the cucumbers and gourds when the Etesians winds have begun to blow, and by this means render them sweeter and tenderer without having to water them."

νει καὶ τὸ σαρκῶδες αὔξει, διὰ ταῦτα καὶ κατὰ λόγον ἀμφοῖν τὸ συμβαῖνον. ποιεῖ δὲ καὶ εὐχυλότερα καὶ μείζω καὶ ἡδίω μετὰ τῆς ἄλλης θεραπείας ὅτι καὶ τροφὴ[1] πλείων καὶ πέψις γίνεται μᾶλλον. ἐπὶ δὲ τοῦ κράνου (περὶ τούτου γὰρ μάλιστα ἀντιλέγεται) δῆλον <ὡς> (εἴπερ [ὡς] ἀληθές)[2] πλεῖον[3] ἂν λαμβάνοι[4] τῆς συμμέτρου τροφῆς, ὥσθ' ἧττον ποιεῖν εὔχυλον, ὅπερ καὶ ἐπ' ἄλλων συμβαίνει περικαρπίων ἐν ταῖς πιείραις λίαν χώραις καὶ εὐτρόφοις.

10.3 ἡ μὲν οὖν σκαπάνη πάντα ταῦτα συναπεργάζεται, καὶ τὸ ὅλον ὠφελεῖ, διὰ τὸ ἐξαιρεῖν τὰ παραιρούμενα· ἐπεὶ καὶ τὰ παραφυτευόμενα καὶ τὰ παρασπειρόμενα διὰ τοῦτο βλάπτει πάντα, τὰ δὲ καὶ ὅλως ἀναιρεῖ, πλὴν ὅσα γ' ἐν φαρμάκου μέρει· λέγω δ' ὡς οἱ τὰς κριθὰς ἐπισπείροντες τοῖς τῶν ἀμπέλων φυτοῖς, ἢ εἴ τι ἄλλο ξηρόν, ὅπως

[1] u aP : -φῆς U N.
[2] ego (εἴπερ ὡς ἀληθῶς Schneider) : εἴπερ ὡς ἀληθὲς U.
[3] u : πλείων U.
[4] ego : ἀναλαμβάνοι U (-ει N aP).

[1] *Cf. HP* 3 2. 1: "... and they (*sc.* wild trees) promise more fruit (*sc.* than cultivated ones) but concoct it less, if

the fleshy part grow, and on this account what happens to both parts is reasonable. Spading also (when joined to the other kinds of care) makes the fruit juicier, larger and better tasting because not only is there more food but also better concoction. In the case of the cornel cherry (for it is here that the point is most disputed)[1] it is clear (if the fact is true) that the tree must get more than the right amount of food, so that cultivation makes the fruit less juicy; and this also happens with other pericarpia in countries that are fat and nutritious to excess.[2]

An Excursus on Good and Bad Neighbours

So spading helps to bring all this about and is generally beneficial because it removes whatever intercepts the food. Indeed plants also that are planted or sown as neighbours are all of them injurious for this reason, some actually destroying a tree, except where they serve as an antidote. I mean the cases where growers sow barley (or some other dry

10.3

not all of them, at least the wild varieties when compared with the cultivated ones of the same tree ... For all such wild varieties do so, with rare exceptions, as with cornel cherries and sorb apples, since they say that here the wild fruit is mellower and sweeter than the cultivated ..."

[2] *Cf. CP* 2 4. 3.

THEOPHRASTUS

τῆς ὑγρότητος ἀφαιρεθῇ, καὶ ὡς ταῖς ῥαφανῖσι[1] τοὺς ὀρόβους πρὸς τὸ[2] μὴ κατεσθίεσθαι,[3] καὶ εἰ δή
10.4 τι ἄλλο τοιοῦτον· καὶ ὅσα προσφιλῆ τυγχάνει, καθάπερ δοκεῖ τῶν δένδρων ἐλάα καὶ μύρρινος· τάς τε γὰρ ῥίζας συμπλέκεσθαί φησι τῶν δένδρων Ἀνδροτίων, καὶ τὰς τοῦ μυρρίνου ῥάβδους φύεσθαι διὰ τῶν ἀκρεμόνων τῆς ἐλάας, τόν τε καρπόν, ἔχοντα τοῦ ἡλίου καὶ τῶν ἀνέμων προβολὴν τὴν ἐλάαν, ἁπαλὸν[4] γίνεσθαι καὶ γλυκύν, ἐλάττω[5] δὲ τοῦ ἐν τοῖς προσείλοις. ἄλλο δ' οὐδὲν ἐθέλει φυ-
10.5 τεύεσθαι πρὸς ταῖς ἐλαίαις. εὐμενὲς δὲ καὶ ἡ πεύκη δοκεῖ πᾶσιν εἶναι, διά τε τὸ μονόρριζος εἶναι καὶ βαθύρριζος· ὑποφυτεύεται γὰρ ὑπ' αὐτῇ καὶ μύρρινος καὶ δάφνη καὶ ἕτερα πλείω, καὶ αὔξησιν λαμβάνει. καὶ φανερὸν ὅτι μᾶλλον παροχλοῦσιν αἱ ῥίζαι τῆς σκιᾶς·[6] σκιὰν γὰρ πολλὴν ποιεῖ [ἡ πεύκη].[7] κατὰ λόγον δὲ καὶ τἆλλα τὰ ὀλιγόρριζα καὶ βαθύρριζα, καὶ πρὸς τοῖς[8] ὧν[9] μὴ κατὰ τὴν αὐτὴν
10.6 ὥραν ἡ βλάστησις καὶ καρποτοκία.[10] τροφήν τε γὰρ ἐλαχίστην παραιρεῖται ταῦτα, καὶ παραλλάτ-

---

[1] Schneider : ῥαφάνοις U. [2] U aP : τῷ u N.
[3] Heinsius : κατατίθεσθαι U.
[4] u P : ἁπαλὴν U (ἁ- N a).
[5] Schneider after Gaza : ἔλαττον U.

plant) among vine slips to reduce their wetness,[1] or sow bitter vetch among radishes to save them from being devoured,[2] and the like. Another exception are the companionable plants, as olive and myrtle are held to be among trees. So Androtion says that they entwine their roots together and that the canes of the myrtle grow between the branches of the olive, and its fruit, screened by the olive from sun and wind, becomes tender and sweet, although there is less of it than in the trees that stand in the sun. But nothing else cares to be planted next to the olive. The pine too is held to be friendly to all because it has a single root and the root goes deep. So myrtle, bay and a number of other trees are planted under it and grow in size. And it is clear that roots interfere more than shade, since the pine casts a large shadow. Other trees are correspondingly friendly as their roots are fewer and deeper; and so moreover are trees with different times of sprouting and fruiting. For all these intercept the smallest amount of food and their different season

10.4

10.5

10.6

---

[1] *Cf. CP* 2 18. 1 and note *a*.
[2] *Cf. CP* 2 18. 1 and note *b*.

---

6 Gaza (*quam umbram*), Itali : ταῖς σκίλλαις U.
7 ego.
8 U : τούτοις Wimmer.
9 Heinsius : ᾧ U.
10 U^r N aP : -αν U^{ar}.

THEOPHRASTUS

τοντες οἱ χρόνοι τῇ τε βλαστήσει καὶ τῇ τελειώσει τῶν καρπῶν ἧττον λυποῦσι. χαλεπώτατα δὲ καὶ ἀμπέλῳ καὶ τοῖς ἄλλοις συκῆ καὶ ἐλάα· καὶ γὰρ τροφὴν πολλὴν ἀμφότερα λαμβάνει, καὶ σκιὰν παρέχει πλείστην· χαλεπὸν δὲ καὶ ἡ ἀμυγδαλῆ, διά τε τὴν ἰσχὺν καὶ διὰ τὴν πολυρριζίαν. καίτοι φυτεύουσί τινες ἐν ταῖς ἀμπέλοις ὡς ἀσινέστατον, τροφῆς τε ἐλαφρᾶς δεομένην καὶ ἅμα διὰ τὸ πρωΐκαρπον καὶ μὴ πολύσκιον ἥκιστα ἐνοχλοῦσαν· οὐ

10.7 μὴν ὀρθῶς γε[1] λέγουσιν. οὐ γὰρ οὕτως ἡ σκιὰ λυπεῖ (καθάπερ εἴρηται) καὶ τὸ ὄψιον ὡς ἡ ἰσχὺς τῶν ῥιζῶν· ἀφαιρεῖται γὰρ τὴν τροφὴν σύνταρρον γινόμενον.[2] ἀλλὰ κουφότατα[3] καὶ ἀσινέστατα[4] πάντων ἐστὶ μηλέα καὶ ῥόα· καὶ γὰρ οὐ πολύρριζα, καὶ τροφῆς ἐλαφρᾶς δεῖται, καὶ ταχὺ γηράσκουσιν, ὥστε μὴ πολὺν χρόνον ἐνοχλεῖν. ἁπλῶς δὲ πάντ᾽[5] ἐπισινῆ καὶ βλάπτει τῇ παραφύσει.

10.8 καίτοι συμβαίνειν[6] γ᾽ ἐνίοτε, τῶν παραφύτων[7] παραιρουμένου θατέρου, καὶ θάτερον αὐαίνεσθαι· τοῦτο γὰρ ἤδη τινὶ συνέπεσεν, ἅμα πεφυκυιῶν ἀναδενδράδων καὶ συκῶν, ὡς αἱ συκαῖ παρῃρέθη-

[1] aP : τε U N.   [2] U : συντάρρων γινομένων Schneider.
[3] Gaza (facillime [-ae G^{ur}]) : κουφότατον U.   [4] U : -ον u.
[5] ego (tamen omnia Gaza, δ᾽ ἅπαντα Itali) : δ᾽ πάντων U : δ᾽ ἁπάντων N aP.

makes them less harmful both to sprouting and to maturing the fruit. The worst neighbours not only for the vine but for the rest are fig and olive, since they both take a great amount of food and cast a very large shade. The almond too is a bad neighbour both because of its strength and because of its many roots. And yet some growers plant it among the vines as being very harmless, needing but light food and then too interfering least of all, since it fruits early and does not cast a great shade. But their reasoning is mistaken. For it is not shade (as we said)[1] and fruiting late that does the harm so much as the strength of the roots, since the tree intercepts the food when its roots become matted with those of the other. The easiest and most harmless neighbours of all are rather the apple and pomegranate, since they are not many-rooted, need but light food, and rapidly age (and so are not a burden for long). But absolutely speaking all trees are harmful and do injury by growing near.

10.7

Yet we hear that it sometimes occurs that when one of two neighbours is removed the other withers. For this once happened to a man who had fig-trees growing with his climbing grapes after he took away

10.8

---

[1] *CP* 3 10. 5.

[6] U ar : -ει U r N aP.
[7] Schneider : παραφυτῶν U.

THEOPHRASTUS

σαν. αἴτιον δ' ὅτι μεγάλην ἐποίησεν μεταβολήν, συνηυξημένων ἤδη καὶ συνεκτεθραμμένων, ὥσπερ γὰρ μία φύσις ἐγεγένητο διὰ τὸν χρόνον· εἰ δὲ νέων[1] ὄντων εὐθὺς παρεῖλεν, οὐχ ὅτι ἂν ἀφήυανεν,[2] ἀλλὰ καὶ εὐαξεστέρας καὶ καλλίους ἐποίησεν.

καὶ περὶ μὲν τῶν ἄλλων δένδρων ἐκ τούτων θεωρείσθω.

11.1 περὶ δὲ ἀμπέλων ὅσα μὴ κοινὰ καὶ ἐν τοῖς πρότερον εἴρηται λεκτέον ὁμοίως.

ἐπεὶ δὲ καὶ τὰ γένη διαφέρει, καὶ αἱ χῶραι, τοῦτο χρὴ πειρᾶσθαι διαιρεῖν· τὰ ποῖα, καὶ[3] ταῖς ποίαις, οἰκεῖα· κατὰ φύσιν μὲν γὰρ ἐὰν φυτεύῃ τις, ἀγαθά, παρὰ φύσιν δέ, ἄκαρπα γίνεται. τὸ δὲ

---

[1] ego (*nam si cum novellae ... essent* Gaza : εἰ μὲν νέων Itali : ἐπεὶ εἴ γε νέων Schneider : ἐπεὶ νέων Wimmer) : εἰνένεων U.
[2] ego (ἂν ἀφαύανεν Wimmer) : ἀναφηανεν U.
[3] [καὶ] Gaza, Scaliger.

---

[1] For the whole subject *cf.* G. Senn, "Der Rebbau im antiken Griechenland," *Gesnerus*, vol. i (1944), pp. 77–91.
[2] *Cf. HP* 2 5. 7: "For when vines are planted naturally

the trees. The reason is that he made too great a change, the vines having by then grown to the tops of the trees and been brought up in their society, for a single nature (as it were) had come about from the long time spent together. But if he had removed the trees when the vines were still young, far from making them wither, he would have made them faster-growing and finer vines.

The other trees are to be studied in the light of this discussion.

The Vine: Special Measures

As for the vine, we must similarly discuss measures not common to the rest and not treated earlier.[1]

Matching the Kinds of Vine with the Kinds of Country

Since not only the varieties of vine but also the countries differ, we must endeavour to distinguish what varieties are appropriate to what countries, since if one plants them in the natural way they turn out well, whereas if one plants them unnaturally they fail to bear.[2] Authorities find what is

---

they are said to turn out well, but when planted unnaturally they fail to bear."

THEOPHRASTUS

κατὰ φύσιν σχεδὸν καθ' ὁμοιότητά τινα λαμβάνουσιν (ὥσπερ εἴπομεν), ἐν μὲν τῇ στερεᾷ καὶ αὐχμώδει τὰ στερεά, καὶ τῶν λευκῶν καὶ τῶν μελάνων[1] (ὡς δ' ἐπὶ τὸ πᾶν τὰ μέλανα στερεώτερα), ἐν δὲ τῇ ἐπόμβρῳ τὰ μανά. διάδηλα δὲ τὰ πυκνὰ καὶ τὰ μανὰ ταῖς μήτραις, ἃς δεῖ[2] θεωρεῖν ἀποτέμνοντας τὰ νέα τῶν κλημάτων· ἔχει γὰρ ἡ μὲν μανὴ πολλήν, τὸ δὲ ξύλον λεπτόν, ἡ δὲ πυκνὴ μικράν, τὸ δὲ ξύλον παχύ.

11.2 ὅτι δὲ ἑκάτερον ἑκατέρᾳ τῇ χώρᾳ συμφέρει, διὰ τῶνδε φανερόν· ὧν μὲν γὰρ ἡ μήτρα μεγάλη, τούτων καὶ πολλοὶ πόροι[3] καὶ[4] εὐρέες, ὧν δὲ μικρά, στενοί τε καὶ ὀλίγοι, διὸ καὶ αὐτὰ στερεά·[5] τροφῆς <δ'>[6] ἐλάττονος δεῖται τὰ πυκνά, καὶ ἅμα διατηρεῖν δύναται τὸ ὑγρὸν εἰς τὸ θέρος· ὁ γὰρ ἥλιος οὐχ ὁμοίως ἐξάγει διὰ τὴν πυκνότητα, διὸ μικρᾶς τῆς ἀφαιρέσεως γινομένης ἐπαρκοῦσιν αἱ ῥίζαι τὸ σύμμετρον εἰς τὸν καρπόν· αἱ δὲ μαναί, πολλῆς μὲν τῆς ἀφαιρέσεως οὔσης, ὀλίγης δὲ τῆς ἐπιρροῆς διὰ τὴν ξηρότητα τῆς γῆς, τά τε

---

[1] u : μελανῶν U : μελαινῶν φυτεύοντες Schneider.
[2] u aP : δὴ U N.
[3] ego (οἱ πόροι <πολλοὶ> Wimmer : *meatus ampli multique* Gaza) : οἱ πόροι U.

DE CAUSIS PLANTARUM III

natural (as we said)[1] by following a certain similarity (as it were): in firm and rainless country they plant the vines of solid texture (both of the white and dark varieties; but on the whole the dark are the more solid), in rainy country the vines of open texture. Close and open texture can be distinguished by the types of pith, which must be studied in cross sections taken from the younger branches: the open-textured vine has more pith than wood, the close-textured more wood than pith.

That each of the two types of vine thus distinguished goes with the type of country described is clear from the following: where the pith is large the passages are numerous and wide; where it is small they are narrow and few and the vines are therefore themselves solid. Close textured plants require less food and then too are able to retain their fluid for the summer, since the sun does not extract so much of it because of the closeness of their texture. Consequently, the loss being small, the roots supply the right amount for developing the fruit. The vines of open texture, on the other hand, the loss being great and the influx small (owing to the dryness of the

11.2

---

[1] *CP* 2 4. 7.

[4] [καὶ] u aP.

[5] αὐτὰ στερεά ego (Gaza omits : Schneider deletes : διὰ τοῦ ἔαρος Wimmer) : αὐτὰ στερεᾶς U.

[6] ego.

κλήματα ἀσθενῆ καὶ τοὺς καρποὺς ἀτελεῖς φέρουσιν. τῇ δὲ ἐπόμβρῳ προσφορώτατον τὸ γένος, ἅτε
πολλῆς τροφῆς δεόμενον· καὶ μεγάλης ἀφαιρέσεως γινομένης (ὁ <γὰρ>[1] ἥλιος ἐξάγει), τοσούτῳ πλέον[2] ἐπιδίδωσιν καὶ εἰς μέγεθος καὶ εἰς
εὐκαρπίαν, ἄφθονον μὲν τροφὴν ἔχουσα,[3] ταύτης
δὲ κατακρατοῦσα καὶ πέττουσα ῥᾳδίως. διὰ τοῦτο
γὰρ καὶ ἡ λειμωνία δοκεῖ ταῖς ἀμπέλοις εἶναι καὶ
κρατίστη, διότι πρὸς τῷ[4] κούφη καὶ μὴ πίειρα εἶναι, καὶ ὕφυδρός[5] ἐστιν ὥστε μὴ δύνασθαι τὸ οὐ-
ράνιον ὕδωρ δικνεῖσθαι πρὸς τὸ ἐκ τῆς γῆς. ἡ δ᾽
ἄμπελος ὕδατος πλείστου δεῖται διὰ τὸ καὶ ἐν τῷ
καρπῷ τὸ[6] πλεῖστον ἔχειν τὸ ὑγρόν· ἔτι δὲ
μάλιστ᾽ ἀντέχειν δύναται ἐν ταῖς ἐπομβρίαις.

ἐὰν δὲ ἡ χώρα μήτε αὐχμηρὰ μήτε ἔπομβρος,
ἀλλὰ μέση τυγχάνῃ, τὰ ἀνὰ μέσον φυτεύειν, ὅσα
μήτε πυκνὰ μήτε μανά.

γένη μὲν οὖν πρὸς ἑκάστην ταῦτα οἰκεῖα διαι-
ρεῖται.

---

[1] ὁ <γὰρ> ego (quo ... magis Gaza : <ὅσῳ πλέον> ὁ Schneider.
[2] U : μᾶλλον Schneider.
[3] Wimmer : ἔχει U.   [4] τῷ : το U.
[5] Schneider (cf. CP 2 4. 4) : ἔφυδρος U.
[6] [τὸ] aP.

ground),[1] bear weak branches and unfinished fruit. But this variety is very well adapted to rainy country, needing as it does a great deal of food; and although the vine loses a great deal of fluid (for the sun extracts it),[2] it increases all the more both in size and in excellence of fruit, since it has an unstinted supply of food and masters and concocts it easily. This in fact is why meadow land is regarded as the very best for vines: besides being light and not fat, it also contains water and so prevents the rain water from joining the water coming from the earth.[3] And the vine requires the greatest amount of water because it is also the tree with fluid as the largest component of its fruit; and it is further the tree best able to hold out in rainy weather.

11.3

11.4

If the country is neither droughty nor rainy but intermediate, we are told to plant the intermediate varieties, neither close nor open in texture.

So the kinds of vine that are distinguished as appropriate to the different countries are these.

---

[1] That is, of the ground suited to the solid vine.

[2] From the open texture.

[3] *Cf. CP* 2 4. 4; *Geoponica*, v. 1–2: "Black earth, which is not tightly packed, and which has below ground a moderate supply of sweet water, is suited to the vine. For earth like this, when it receives the rain, neither loses it by sending it below entirely, nor keeps it on the surface above ..."; Columella, *On Trees*, chap. iii. 7 (on land good for planting the vine): "but the land which transmits the rain or keeps it long on the surface is to be avoided."

THEOPHRASTUS

11.5 περὶ δὲ τῶν σπερμάτων, ἐπείπερ ἰσχυρότατα δεῖ, διὰ τοῦτό τινες κελεύουσιν ὡς ἐκ ψυχροτάτης χώρας λαμβάνειν · πυκνότατα γὰρ ὄντα μᾶλλον ἔχειν[1] ὥστε καὶ ἐν ταῖς λεπταῖς ἀντιλαμβάνεσθαι, καὶ ἐν ταῖς ἐπομβρίοις[2] μὴ ἐκσήπεσθαι (διὰ τοῦτο γὰρ καὶ τὰ μοσχεύματα δεῖν εἰς[3] τὰς ἐπομβρίους μᾶλλον ἐμβάλλειν ἢ τὰ φυτεύματα πάντων τῶν δένδρων, ὅτι τὰς ῥίζας τὰς καθιεμένας τῶν φυτευμάτων, ἀσθενεῖς οὔσας, ἐκσήπει,[4] [ἡ][5] τῶν <δὲ>[6] μεμοσχευμένων ἰσχυρότεραι,[7] καὶ εὐθὺς ἀντιλαμβάνεται).

11.6 τὰ δὲ πάχη τῶν φυτῶν εἰς μὲν τὴν ἔπομβρον οἰκεῖα · δεῖ γὰρ ὅτι[8] μάλιστα ἰσχυρά, καθάπερ λέγομεν · εἰς δὲ τὴν αὐχμηρὰν μήτε παχέα μήτε ἄγαν λεπτά · τὰ μὲν γὰρ οὐκ ἂν δύναιντο σῆψαι, τὰ δὲ ἀσθενῆ κίνδυνος μὴ πρὸ τῆς βλαστήσεως ἀποξηρανθῇ.

παραλλάττουσι δὲ καὶ οἱ χρόνοι τῆς φυτείας καθ' ἑκάτερον ·[9] τὴν μὲν γὰρ ἔπομβρον καὶ ψυ-

---

[1] U : ἀντέχειν Heinsius.
[2] ego : -ίαις U.   [3] Schneider : ἐς U.
[4] ἐκσήπει <ἡ ὑγρότης> Schneider after Gaza.
[5] N^{ac} M (ἡ U : N^c adds ἡ after τῶν) : ἢ aP.
[6] Schneider after Gaza.   [7] U^c from -α.
[8] U^{cm} (with index) : U^t omits.
[9] U : ἑκατέραν Schneider.

DE CAUSIS PLANTARUM III

The Vine: Planting

As for the vine slips, since they should be very strong we are told by some agriculturists to take them from the coldest possible country, since when the slips are closest in texture they are better able (they say) not only to take hold in thin country but also not to rot out in rainy country. (This indeed is why we are told with all trees to plant rooted slips[1] and not just cuttings in country with heavy rain, because the ground rots out the roots that are sent down by a cutting, since they are weak, whereas the roots of a rooted slip are stronger and the slip takes hold at once.) 11.5

Thickness in a slip suits it for rainy country, since (as we are saying)[2] the slip should be the strongest possible; but for ground with little rain neither a thick nor a very slender slip is suitable, since the rainfall would never be able to decompose[3] the thick ones, and the weak ones are in danger of drying out before they can sprout. 11.6

The time of planting also differs for the two regions; so one should plant rainy and cold country

---

[1] "Rooted slips" renders *moscheumata*: for this see note 1 on *CP* 3 5. 3.

[2] *CP* 3 11. 5.

[3] That is, make them germinate; the usage was doubtless taken from an agricultural writer.

THEOPHRASTUS

χράν ὀλίγον πρὸ ἰσημερίας δεῖ, τότε γὰρ ξηρότατα καὶ θερμότατα·[1] τὴν <δ'>[2] αὐχμηρὰν καὶ θερμὴν μεθ' ἡλίου τροπάς· ὅσῳ γὰρ ἂν μᾶλλον βρέχηται, τοσούτῳ βελτίων ἡ βλάστησις· τὴν δ' ἄλλην ἅπασαν τεκμαιρόμενον πρὸς τοῦτο τὸν[3] χρόνον.

12.1   χρὴ δὲ καὶ τὴν ἐργασίαν ποιεῖσθαι πρὸς τὴν χώραν, εὐθὺς ἐπ'[4] αὐτῶν τῶν γύρων ἀρξαμένους· οἷον ἐν τῇ ἐπόμβρῳ μήτε μεγάλους ὀρύττοντας μήτε βαθεῖς, ὅπως μὴ πολὺ συνιστάμενον ἐκσήπῃ τὸ ὕδωρ (διὰ τοῦτο γὰρ ἐὰν σφόδρα κάτομβρος ᾖ τοῖς παττάλοις τοῖς σιδηροῖς[5] φυτεύουσιν)· μηδὲ δὴ συσκάπτειν αὐτοετεί,[6] μηδὲ τῷ[7] ὕστερον, ὅπως μάλισθ' ὁ ἥλιος ξηραίνῃ τὴν κατὰ τὸ φυτὸν γῆν· ἐὰν δὲ αὐχμηρὰ[8] καὶ ξηρά,[9] δῆλον ὡς ἐναντίως· οὐ γὰρ γύρους, ἀλλὰ[10] πᾶσαν, εἰ δυνατόν, ὀρυκτέον ἵν' ὡς μάλιστα συνεργασθεῖσα δέχηται τὸ ὕδωρ. εἰ δὲ μή, τάφρους ὡς βαθυτάτας καὶ μεγίστας ποιοῦντα[11] συσκάπτειν ὅτι μάλιστα

[1] U : ξηροτάτη καὶ θερμοτάτη Gaza, Schneider.
[2] aP.   [3] Wimmer : τοῦτον τὸν a : τοῦτον U N P.
[4] U : ἀπ' N aP.
[5] u : σιδήροις U.
[6] Gaza (anno eodem), Schneider : αὐτὸ, ἔτι U.
[7] τῷ U ar : τὸ U r N aP.

shortly before the equinox,[1] since the slips will then be driest and warmest, but arid and hot country after the solstice,[2] since the more rain the slip receives the better it sprouts. One should plant all other kinds of country by figuring out the time from these two cases.

One must also suit the working of the soil to the country, starting with the holes themselves. So in rainy country we must not dig the holes large or deep, to keep rain from collecting in large amounts and rotting the slips out (for this is why if the country is very rainy indeed the hole is made by driving an iron peg into the ground). Nor should we spade the ground into a mound round the plant in the same year nor yet in the next, to let the sun dry the earth round the slip as much as possible. But if the country has little rain and is dry we must evidently do the opposite, not digging mere holes but if possible the whole ground, so that after this working over it may best absorb the rain. Failing that, we should dig the deepest and biggest trenches we can and then spade the ground into mounds in as close a contact with the plant as we can manage, to keep it

12.1

---

[1] Of autumn.
[2] Of winter.

[8] u : αὐχμηρὰν U N aP.   [9] U N : ξηρὰν aP.
[10] Wimmer (ἀλλ' ἅμα Schneider) : ἅμα U.
[11] U N P : ποιοῦντας a.

THEOPHRASTUS

πρὸς αὐτὸ τὸ φυτόν, ὅπως ἥκιστα διαθερμαίνηται τοῦ θέρους.

12.2 ὁμοίως δὲ καὶ τἆλλα τὰ τούτοις ἀκόλουθα ποιεῖν. ὡς γὰρ ἁπλῶς εἰπεῖν ἀληθὲς τὸ καὶ πρότερον λεχθέν, ὅτι δεῖ ταῖς ἐργασίαις τὴν μὲν ὑγρὰν ξηραίνειν, τὴν δὲ ξηρὰν ὑγραίνειν· διὸ καὶ τοῦ ἔαρος[1] κελεύουσιν ὡς βαθύτατα σκάπτειν, ὅπως ὅτι πλεῖστον ἐγκαταμιχθὲν πνεῦμα παράσχῃ[2] τροφήν· ὥσπερ γὰρ εἴρηται πολλάκις, οὐκ ἀπὸ τῆς γῆς μόνον, ἀλλὰ καὶ ἀπὸ τοῦ ἡλίου καὶ ἀπὸ τοῦ ἀέρος ἡ τροφή.

12.3 λαμβάνουσι δὲ καὶ κατὰ τὰς ὀρεινὰς καὶ ἀνεμώδεις τὰ πρόσφορα ταῖς φυτείαις, οἷον τούς τε γύρους ὀρύττοντες τοῖς καθεστῶσι πνεύμασι καὶ τὸ φυτὸν τιθέντες τὸν αὐτὸν τρόπον, ὅπως ἡ βλάστησις μὴ ἐναντία γίνηται τῷ πνεύματι, θραύεσθαι γὰρ ἐκείνως ἀνάγκη, μηδ' ἔμπαλιν ὑπερπεσοῦσα,[3] παρὰ φύσιν αὔξηται· τούτων γάρ τι πάσχουσα χείρων ἡ ἄμπελος. ὡσαύτως δὲ καὶ τὰ ἄλλα τὰ ἀκόλουθα πρὸς τὴν χώραν.

---

[1] u Gaza a : ἀέρος U N P.
[2] Gaza (*praestare*), Heinsius : κατάσχῃι U.
[3] U : ὑποπεσοῦσα Schneider.

DE CAUSIS PLANTARUM III

as far as possible from getting overheated in summer.

12.2 In the same way we must also carry out all the operations that follow these. For in a word what was said before[1] is true, that in working the earth one should make wet ground dry and dry ground wet. This is why the experts tell us to work the ground as deep as we can in spring, so that as much *pneuma*[2] as possible may be mixed in with it and provide food, since food (as we have often said)[3] comes not only from the earth but also from the sun and from the air.

12.3 The experts also consider the mountainous and windy countries in taking the proper measures in planting. For example they dig the holes during the prevailing wind, placing the slip in the same position to it that it had on the vine,[4] so that it may not sprout against the prevailing wind, since in that case breakage is inevitable, nor yet be blown over in the other direction and grow unnaturally, for the vine affected in any of these ways is inferior. So too they carry out the procedures that come next with due regard to the country.

---

[1] *CP* 3 10. 1.
[2] That is, warm air containing water and leading to expansion; *cf. CP* 3 10. 1, where the same view is cited, but with "air" replacing *pneuma*.
[3] *CP* 3 2. 1; 3 4. 2, 3.
[4] *Cf. CP* 3 5. 2.

THEOPHRASTUS

τὰς μὲν οὖν φυτείας ἐν ἑκάστοις οὕτω καὶ διὰ ταῦτα ποιοῦνται.[1]

13.1 τὰς δὲ τομὰς πανταχοῦ τῶν φυτῶν βραχείας ὡς ἂν ῥιζωθῶσιν καὶ αὐξηθῶσιν· ἅμα δὲ ἡ αὐχμηρὰ καὶ τρέφειν ἀδύνατος τὰ πολλά. κατὰ δὲ τὴν ὀρεινὴν καὶ ψυχράν, ᾗ καὶ ὅλως ἔναυρον,[2] καὶ τὸ ἔαρ τέμνειν εὐλαβοῦνται, τὸ γὰρ ψῦχος ἀποξηραίνειν.[3] ὥραν δὲ τῆς τομῆς σχεδὸν τὴν αὐτὴν πᾶσιν ἀποδιδόασιν, ὑπ' αὐτὴν τὴν βλάστησιν, ὅπως τὸ φυτόν, πλῆρες ὂν τοῦ ὑγροῦ, πρὸς τὴν ἀφαίρεσιν ὁρμήσῃ, καὶ ὁ βλαστὸς ὅτι κάλλιστος γένηται·
13.2 τὴν γὰρ τομὴν ταύτην οὐ καρποῦ χάριν, ἀλλὰ βλαστοῦ γίνεσθαι.

τὴν δὲ μετοπωρινὴν καρποῦ, διόπερ ἐκείνην ποιητέον εὐθὺ μετὰ Πλειάδος δύσιν· τότε γὰρ εἶναι συνεστηκότα τε μάλιστα, καὶ ἥκιστα τεμνόμενα δακρυρροεῖν καὶ ῥήγνυσθαι. μεθ' ἡλίου δὲ

---
[1] a : -τα U N P.
[2] Schneider (cf. HP 8 11. 6) : ἔναστρον U. [3] ego : -ει U.

---
[1] Cf. CP 3 15. 5. [2] In summer.
[3] Cf. CP 3 13. 3 (training the vine to grow evenly around the stock makes it bear better fruit).

DE CAUSIS PLANTARUM III

So the planting of slips in the various regions is carried out as described and for the reasons given.

The Young Vine: Cutting Back

The slip is everywhere cut short to let it root and grow[1]; then too when there is no rain[2] the earth is unable to feed the greater part of the shoots. In uplands and cold country, and in general where there is much breeze, cutting back is avoided even in spring, since the cold (they say) dries the slips out. But the authorities give the same season (one may say) for all countries, just at the time of sprouting, so that the slip, full of its fluid, may in its impulse to grow take the direction given by the removal, and the resulting shoot may be the finest possible. For this cutting back (they say) is not for fruit but for sprouts.

13.1

13.2

The Young Vine: Training

Cutting in autumn, on the other hand, is (they say) for fruit,[3] which is why it must be carried out directly after the setting of the Pleiades, since then the young vines are firmest[4] and least apt to bleed and crack[5] when cut, whereas after the solstice[6]

---

[4] That is, least fluid and watery, since the rains have not yet begun.

[5] *Cf. CP* 3 7. 10 for other trees.   [6] Of winter.

THEOPHRASTUS

τροπὰς καὶ μετὰ ζεφύρου πνοὰς ἀμφότερα πάσχειν[1] ταῦτα τεμνόμενα, καὶ τό τε φυτὸν πο-
νεῖν,[2] καὶ τὸν βλαστὸν ἐξ ἄκρου φύεσθαι πρὸς τῇ
τομῇ, διὰ τὸ τὴν τροφὴν τὴν μὲν ἐξερρυηκέναι,
τὴν δ' ἐν τοῖς ἄκροις καταλελεῖφθαι.

13.3 τέμνειν δὲ τὰ μὲν ἐν τῇ χέρσῳ φυτευόμενα τῷ
τρίτῳ ἔτει, θᾶττον γὰρ παραγίνεσθαι διὰ τὸ νεορ-
γοτάτην εἶναι τὴν γῆν καὶ ἀκάρπωτον· τὰ δ' ἐν
τῇ γεωργουμένῃ ὀψιαίτερον.

τὴν δ' ἄμπελον ἄγειν δεῖ κύκλῳ περὶ τὸν πυθ-
μένα, πανταχόθεν γὰρ ὁμαλοῦς οὔσης καλλίων καὶ
εὐκαρποτέρα·[3] τοῦτο δ' οὐ χαλεπόν, ἐάν τις
καταλίπῃ μὴ τὰ κάλλισθ' ὡρμηκότα τῶν κλημάτων, ἀλλὰ τὰ ἄριστα πεφυκότα πρὸς τὴν ἀγω-
γήν· ἀρκεῖ γὰρ εἷς ὀφθαλμὸς λειφθεὶς[4] εἰς τὸ
13.4 δέον. ἐὰν δέ τις ἀπορῇ καὶ ταῦτα,[5] διὰ <τὸ>[6]
τὴν ἄμπελον ὡρμηκέναι κατὰ τὸ αὐτό, τῶν λειφ-
θέντων κλημάτων ἀφελεῖν τοὺς ἐντὸς ὀφθαλμούς,
ὅπως εἰς τοὺς[7] ἔξω ῥέουσα ἡ τροφὴ σχίσῃ τὴν ἄμ-
πελον· ἀεὶ γὰρ πρὸς τὸ ζῶν καὶ τὸ δεχόμενον ἡ
ἐπιρροή, διὸ (καθάπερ ἐλέχθη) ῥᾴδιον ποιεῖσθαι
τὰς ἀγωγάς.

---

[1] U^{ar} : πάσχει U^r N aP.   [2] U^{ar} : πονεῖ U^r N P (ποιεῖ a).
[3] u aP : εὐκαρπότερα U N.   [4] aP : ληφθεῖς U (-εὶς u N).
[5] ego (τοῦτο Gaza, Scaliger) : ταῦτ' οὐ U.

and after the west wind begins to blow[1] the vines (they say) do both when cut, and not only does the plant suffer but the new shoot comes out from the tip of the stump next to the cut because some of the food has bled away and the rest is left at the tip.

In virgin soil one should cut in the third year after planting, since here they say the vines grow up faster because the soil is the most recently worked there is and no crop has been taken out of it. But in ground under regular cultivation one should cut later.

One must train the vine to grow on all sides of the stock, since when the growth is in an even periphery the vine is finer and bears better. This is not difficult to do if one leaves not the branches that have displayed the greatest vigour but those that grow most conveniently for this training, since a single bud left in the right place is enough. If one cannot even contrive to do this because the vine has taken a single direction in its growth, one must remove the inside buds on the branches that are spared so that the food may flow to the outside buds and make the vine fan out, since the flow is always to the part that is alive and can receive it, which is why training is easy (as we said).[2]

13.3

13.4

---

[1] About February 2; see the Introduction, vol. I p. li.
[2] *CP* 3 13. 3.

---

[6] Scaliger.   [7] U^c : τo U^{ac}.

THEOPHRASTUS

τοῦτο δὲ ἐν ταῖς ἐνίκμοις καὶ ἀγαθαῖς· ἐν δὲ ταῖς αὐχμώδεσι καὶ ξηραῖς ἀνάγκη τὰ βέλτιστα λιπεῖν·[1] ξηρὰ γὰρ οὖσα καὶ ὀλιγότροφος[2] οὐ ῥᾳδίως ἕτερα[3] προήσεται, διὸ συνακολουθητέον τῇ ὁρμῇ· χρὴ δὲ καὶ πρὸς τὰ πνεύματα τῇ αὐτῇ τομῇ χρῆσθαι καὶ μὴ βιάζεσθαι παρὰ φύσιν.

τῶν μὲν οὖν φυτῶν τοιαύτας[4] καὶ διὰ ταῦτα[5] ποιοῦνται τὰς θεραπείας.

14.1 τῶν δ' ἀμπέλων τῶν τελέων ἤδη πρῶτον μὲν καὶ μέγιστόν ἐστιν ἡ κλάσις, καλῶς γὰρ ἀμπελουργουμένη, καὶ εὐβλαστοτέρα καὶ εὐκαρποτέρα καὶ πολυχρονιωτέρα γίνεται· δεύτερον δέ, καὶ τρόπον τινὰ τούτῳ παραπλήσιον, ἡ βλαστολογία, καὶ γὰρ[6] ἐνταῦθα εἰδέναι δεῖ τὰ[7] ποῖα συμφέρει καταλιπεῖν καὶ τὰ ποῖα ἀφαιρεῖν, καὶ πρὸς τοὺς ἑτέρους[8] καρποὺς καὶ πρὸς τὴν ὅλην φύσιν. τὰ δ'

---

[1] N aP : λειπεῖν U : λείπειν u.
[2] U : ὀλιγοτρόφος Schneider.
[3] Gaza (*alios* [-*io* a misprint in G^{ed}]), Schneider : ἑτέρας U.
[4] Gaza (*ad hunc modum*), Schneider : τοιαῦτα U.
[5] aP : ταύτας U N.
[6] U : καὶ γὰρ <καὶ> Gaza (*enim vero ... etiam*), Scaliger.
[7] aP : τὸ U N.
[8] U : Gaza omits : Schneider deletes : ἐτείους Wimmer.

This applies to country that is damp and good. In country that gets little rain and that is dry we are compelled to spare the best branches, since the vine, being there dry and getting little food, does not find it easy to send out new branches, and we must acquiesce in the direction taken by the impulse in the vine. So too we must follow the direction of the wind[1] in cutting and not try to force the nature of things.

Such then is the care bestowed on the young plants and these are the reasons for it.

The Full-Grown Vine: Pruning

When the vines have reached the stage of being full-grown the first and most important step is to prune[2] them, since the vine when properly dressed produces better sprouts and fruit and lives longer. The second step, and in a way a repetition[3] of this, is thinning,[4] since here too we must know what parts are best spared and what are best removed, both for the effect on the rest of the fruiting branches and on the whole nature of the tree. When

14.1

---

[1] So in planting: *cf. CP* 3 12. 3.
[2] *Klásis*, literally "breakage," as contrasted with the schooling or training of the young vine.
[3] *Cf. CP* 3 16. 1.
[4] *Blastologia*, literally "selecting the shoots," is the pruning away of fruiting shoots; *cf. CP* 3 16. 1.

THEOPHRASTUS

ἄλλα ἤδη κοινότερα καὶ ῥάω, πάντα δ' ἔχοντα διάνοιαν,[1] ὥστε καιρὸν ζητεῖν καὶ τρόπον.

14.2 αἱ δὲ διαφοραὶ κατὰ τὰς χώρας[2] τῶν ἔργων τὰ[3] μὲν καὶ ἐν τοῖς ἄλλοις εἰσίν, οὐ μὴν ἀλλὰ πλείστη γε κατὰ τὴν κλάσιν καὶ τὴν ἀμπελουργίαν, ὑπὲρ ἧς καὶ πειρῶνταί γε διαιρεῖν, ἅμα τοῖς τε γένεσιν ποιούμενοι καὶ ταῖς χώραις τὸν ἀφορισμόν· ὁ δ' ἀφορισμὸς ἐν δυοῖν· ἐν ποίαις ἑκάστας[4] χώραις[5] δεῖ, καὶ ἐν τῷ βραχυτομεῖν ἢ μακροτομεῖν (ἐπεὶ τά[6] γ' εἰς τὴν ἀγωγὴν καὶ τὴν ὅλην τομὴν τῶν ἀμπέλων πανταχοῦ τὰ αὐτά).

14.3 σχεδὸν δ' ἔνιοί γε τὸν αὐτὸν ἀποδιδόασιν ἐπί[7] τε τῶν γενῶν καὶ τῆς ὁμοίας χώρας ἀφορισμόν· οἷον τὸ βραχυτομεῖν ἔν τε ταῖς καυσώδεσιν καὶ ξηραῖς, καὶ τῶν ἀμπέλων ὅσαι τοιαῦται τυγχάνουσιν,[8] ἐγκαρπότεραι γὰρ γίνονται διὰ τὸ μᾶλλον δύνασθαι τρέφειν· μακροτομεῖν δὲ <ἐν>[9] ταῖς

---

[1] U : διαφοράς Schneider.   [2] Gaza, Schneider : ὥρας U.
[3] U : αἱ Schneider.
[4] ego (ἕκαστα Schneider) : ἕκασταις U.
[5] Gaza, Schneider : ὥραις U.
[6] aP : ἔπειτα U (ἔπειτά N).   [7] u : ἔπει U.
[8] Schneider : τυγχάνουσαι U.   [9] Schneider.

[1] Cf. Plato, Phaedrus 272 A 4–7.   [2] Cf. CP 3 13. 4–5.

we come to the remaining procedures we find that they apply to other trees as well and are easier to carry out, although all involve taking thought, and so require to be done at the right time[1] and in the right manner.

Pruning: Rules that Take the Country into Account

The kind of country makes a difference in other operations too, but it makes the greatest difference in the pruning and dressing of the vine. About this operation the experts endeavour to draw distinctions by an appeal not only to the different kinds of vine but also to the different kinds of country; and the resulting rules bear on two questions: in what kind of country one is to apply the procedure to the various vines, and whether one is to cut off much or little of the branch. (As for the measures taken for training[2] and cutting back[3] the vine, these are everywhere the same.) 14.2

Some authorities at least give the same rule (one may say) for both the type of vine and the type of country of similar character, to wit that we should prune the branches short in torrid and dry country and with vines of this character, since the vines will yield more fruit because they are then better able to feed it; whereas we should prune the branches long 14.3

[3] *Cf. CP* 3 13. 1 (of cutting back): "But the authorities give the same season ... for all countries ..."

THEOPHRASTUS

ἐναντίαις, καὶ τὰς ἐναντίας, οἷον ἐν τοῖς[1] ἐφύγροις καὶ εὐτρεφέσι,[2] καὶ ὅσαι τῶν ἀμπέλων τοιαῦται· φύσει τε γὰρ καρπιμωτέρας[3] εἶναι καὶ τὸν γινόμενον καρπὸν ἐν ἄκροις μᾶλλον φύεσθαι τοῖς κλήμασιν.

14.4 ἔνιοι δὲ καθόλου περὶ πασῶν διαιροῦσιν οὐκ εἰς τὰς χώρας[4] ἀποβλέποντες ἀλλ' εἰς αὐτὰ τὰ γένη καὶ τὰς διαθέσεις, κελεύοντες σκοπεῖν πρὸς τὴν μήτραν (ὥσπερ ἐπὶ τῶν αὐτῶν εἴπομεν) ἐν τοῖς ἀποτεμνομένοις τῶν νέων κλημάτων. ἐὰν μὲν γὰρ ἔχῃ πολλήν, πολλὰ καὶ βραχέα καταλιπεῖν ὅπως βραχέων μὲν ὄντων[5] δύνηται τρέφειν, ἀπὸ πολλῶν δὲ πολὺς ὁ καρπὸς γίνηται· ἐὰν δὲ ὀλίγην καὶ πολὺ τοῦ ἔνου κλήματος [ἔχουσαν],[6] ἐλάττω μὲν τὸ πλῆθος, μείζω δὲ τὸ μέγεθος· δεῖ[7] δὲ ὅσῳ <ἂν>[8] ἐλάττω τὴν μήτραν, τοσούτῳ μακρότερα λείπειν ἐν ἁπάσαις.

14.5 καθόλου μὲν οὖν, οὕτω τὸ μέγεθος <καὶ>[9] τὴν

[1] U : ταῖς Scaliger.
[2] U : εὐτραφέσι u a : εὐτροφέσι N P.
[3] Wimmer (humidae Gaza : καὶ ὑγροτέρας Itali : καὶ πιοτέρας Scaliger) : καὶ ποτέρας U.
[4] Gaza, Schneider : ὥρας U.   [5] N aP : μενόντων U.
[6] ego (ἔχουσα Schneider : ἔχωσιν Wimmer).

in the opposite kind of country and with vines of the opposite type, to wit in wet localities that are well supplied with food and with vines of this character, since these vines are not only naturally more fruitful but the fruit also tends to grow closer to the tip of the branch.

*Pruning: Rules that
Take no Account of the Country*

Some authorities on the other hand make a general distinction applying to all vines, taking no account of types of country, but merely of the variety or state of the vine itself, and tell us to decide by examining the pith (just as we said[1] was done in distinguishing the same two types) in cross-sections of the younger branches: if the vine has a great deal of pith, we are to prune so as to leave many short branches, short so that the vine may be able to feed the fruit, and many so that much fruit may come from them; but if we find that the vine has little pith and a great deal of last year's wood, we are to reduce the number of branches but increase their length, and in all vines the smaller we find the pith, the longer we must leave the branch. 14.4

The general rule then (they say) is so to adjust 14.5

[1] *CP* 3 11. 1.

---

[7] U : ἀεὶ Schneider. [8] ego. [9] Itali.

βραχύτητα τῶν κλημάτων καθ' ἑκάστην ποιεῖσθαι πρὸς τὴν μήτραν τὴν ἐν τοῖς ἀποτόμοις, ὅπως ἴση τὸ πλάτος ᾖ τῷ[1] περιειληφότι κλήματι· τῆς γὰρ ἀμπέλου τεμνομένης ἐξ ἴσου πρός τε τὴν ἕξιν τὴν ἑαυτῆς καὶ πρὸς τὴν τοῦ καρποῦ φοράν, πολυχρόνιον ἔσεσθαι καὶ ἀγαθὴν διὰ τέλους· ἐὰν δὲ ἡ μήτρα πλέον μέρος κατέχῃ τῆς τομῆς, καρπὸν μὲν γίνεσθαι πολύν, τὰ δὲ κλήματα ἀμενηνά,[2] διὰ τὴν τῆς τροφῆς ἀσθένειαν· ἐὰν δὲ αὖ τὸ[3] κλῆμα τὸ ἐν τῇ τομῇ δυνατὸν ᾖ καὶ νεανικόν, καρπὸν ὀλίγον ἀπὸ μικρᾶς τῆς μήτρας· γίνεσθαι γὰρ τὸν μὲν βότρυν ἀπὸ τῆς μήτρας, τὸ δὲ κλῆμα ἀπὸ τοῦ περιειληφότος κλήματος, δηλοῦν δὲ τὴν ἄμπελον αὐτήν· τὰ γὰρ ἐκ τῶν ἔνων νέα[4] βλαστήματα πάντα ἄκαρπα γίνεσθαι διὰ τὴν μικρότητα τῆς μήτρας, καὶ τὰς νέας τῶν παλαιῶν ἀφορωτέρας διὰ τὴν αὐτὴν αἰτίαν, ὡς ἀπὸ τῆς μήτρας τῆς τε σαρκὸς καὶ τοῦ γιγάρτου[5] γινομένων.[6] ὅπερ οὐδ' ὑπολαμβάνουσιν, οὐδ' ἔοικεν, εἴ-

---

[1] τῷ u : τὸ U.
[2] u aP : ἀμενὴν ἄ U : ἀμενὴν N.
[3] u P (αὖτο U) : αὐτὸ N : τὸ a.
[4] U r : νέων U ar N aP.
[5] Gaza (*vinaceum*), Moldenhawer : τοῦτι γὰρ τοῦ U.
[6] ego : -νου U.

the length or shortness of the branches in each vine to the amount of pith appearing in the cross-section that the width of the pith is the same as that of the surrounding wood[1]; for when the vine is pruned with equal regard both for its own condition[2] and for the production of fruit, it will have a long life and will bear well throughout; if on the other hand the pith occupies more of the section than the wood, the fruit will be abundant, but the branches will turn out puny because of the weakness of their food supply; and if again the wood in the section is powerful and sturdy, there will be little fruit from the pith, which is then too small. For the grape cluster (they say) comes from the pith, whereas the branch comes from the surrounding wood; and the vine itself is the proof. For the new shoots from last year's branches are all without fruit because of the smallness of the pith, and young vines are less fruitful than the old for the same reason. This[3] implies that both flesh and stone come from the pith. But this is no view that people hold, and it is also unlikely, since on

14.6

---

[1] If the rim of wood has half the width of the core there will be three times as much wood as core. To produce an equal amount of wood and core the diameter of the core must be to the diameter of the branch as one to the square root of two.

[2] The good condition of the vine (as distinguished from its fruit) depends on the amount of wood in the branches.

[3] To say that the grape-cluster comes from the pith.

THEOPHRASTUS

περ ἐξαιρεθείσης, ἡ μὲν σὰρξ γίνεται, τὸ δὲ γίγαρτον οὐ γίνεται (πλὴν εἰ ἄρα τῆς¹ ἐνδοτάτω [τῆς]² μήτρας ἐξαιρουμένης). τοῦτο μὲν οὖν ἐπι-

14.7 σκεπτέον · εἰ δ' ὅλως αἱ μείζους ἔχουσαι τὰς μήτρας εὐκαρπότεραι καὶ πολυκαρπότεραι τυγχάνουσιν, οὐδὲν ἂν διαφέροι πρὸς τὰ νῦν.

κοινοτάτην μὲν δὴ³ ταύτην εἶναι πάσαις · ὅσαι δὲ κλήματα μὲν πολλὰ φύουσιν, καρπὸν δ' ὀλίγον, τούτων τὰ μὲν ἐξ ἄκρας τῆς πρώρας⁴ ὡς μακρότατα λείπειν, τὰ δὲ πρὸς αὐτὸ τὸ στέλεχος βραχέα, ὅπως ἀπὸ μὲν τῆς βραχείας τομῆς ἡ ἄμπελος αὔξηται, ἀπὸ δὲ τῶν ἐξ ἄκρου κλημάτων ὁ καρπὸς ἀπὸ μεγάλης τῆς μήτρας <ᾖ> πολύς.⁵

14.8 ὅταν δὲ βλαστάνῃ περιαιρεῖν τὰ ἄλλα πάντα, πλὴν ὅσα καρπὸν ἔχει, τούτων⁶ ἐπικνίζειν τὰς κορυφὰς ἐν αὐταῖς ταῖς οἰνάνθαις, ἵνα μήθ' ἡ ἄμπελος εἰς τοῦτο τὸ κλῆμα ἀφιῇ τὴν αὔξησιν ὅπερ ἀποτέμνεται, ἥ τε περιοῦσα τροφή, συνειληθεῖσα ἐπὶ ταῖς οἰνάνθαις, αὔξῃ τὸν βότρυν · ἀεὶ γὰρ δεῖ⁷ τοῦτο ζητεῖν ἐκ τῆς τομῆς, ὅπως ἥ τε ἄμπελος

¹ ego : τοῦ U ar : τὸ U r.   ² ego.
³ v, Gaza : δεῖ U N aP.
⁴ Schneider : τὰς πρώτας U.
⁵ ego : πολὺς <γένηται> Schneider : πολὺς <ᾖ> Wimmer.
⁶ τούτων <δ'> Wimmer.   ⁷ u : δὴ U.

removal of the pith flesh is produced but no stone[1] (unless we suppose that stonelessness results only when the innermost pith is removed). This point, then, needs investigating; as for the general question whether vines with a greater amount of pith produce better and more abundant fruit, it would make no difference to the matter in hand.

14.7

This, then, they say, is the method of pruning of widest application to all vines; as for vines that grow many branches but little fruit, they tell us to prune them so as to leave the branches growing from the end of the main branch as long as possible, but the ones growing up close to the trunk short, so that the vine may grow large from being pruned short and the fruit from the branches at the tip of the main one may be abundant, coming as it does from a large pith.[2]

When the vine comes out we are told (1) to remove all other branches, but (2) in those that promise fruit to pinch off the tips just above the flowers, so that (1) the vine may not expend its growth on the portion cut away and so that (2) the food that is thus saved may be blocked at the flowers and make the cluster grow. For we must always (they say) look to this in pruning: to strength in the

14.8

---

[1] *Cf. CP* 5 5. 1; 5 6. 13.
[2] The pith of the main branch is larger than that of the smaller branches.

THEOPHRASTUS

ἰσχύσει[1] καὶ ὁ καρπὸς ἔσται πολύς.

αὕτη μὲν δὴ καθόλου τίς ἐστι καὶ[2] κοινὴ πάσαις.

15.1 ὥραν δὲ τῆς ἀμπελουργίας οἱ μὲν ταῖς χώραις μόνον διαιροῦσιν, οἱ δὲ[3] τῶν γενῶν ἐφάπτονται.

κελεύουσι δὲ τὰς μὲν ἐν τῇ ξηρᾷ καὶ θερμῇ πρῴας ἀμπελουργεῖν, ὅταν τάχιστα παύσωνται φυλλοβολοῦσαι, τὰς δ' ἐν τῇ ψυχρᾷ καὶ ἐπόμβρῳ μικρὸν πρὸ τῆς βλαστήσεως· αἱ μὲν γὰρ ἐν τῇ ξηρᾷ, διασῴζουσαι τὸ ὑγρόν, αὐταί τε βελτίους γίγνονται καὶ ὁ καρπὸς ἡδυοινότερος· αἱ δ' ἐν τῇ ψυχρᾷ, μεθεῖσαι τὸ ὕδωρ, αὐταὶ[4] μὲν οὐδὲν χείρους, ὁ δὲ καρπὸς ἀσηπτότερος καὶ ἡδυοινότερος· τὰς δὲ ἐν τῇ μέσῃ καὶ εὐκράτῳ κατὰ τὸ μέσον τῶν ὡρῶν.

15.2 ἔνιοι δὲ ὅλως πρωῒ πάσας κελεύουσιν (ὅπερ[5] εἴπομεν), ὅτι παχύτερόν τε τὸ κλῆμα γίνεται καὶ

[1] P : ἰσχύσηι U (ἰ- u : ἰσχύση N [-ῃ a]).   [2] U : u erases.
[3] δὲ <καὶ> Gaza (etiam), Schneider.
[4] Gaza (ipsae), Schneider : αὗται U.
[5] U : ὥσπερ Schneider.

[1] These authorities are not discussed until CP 3 15. 4.
[2] Since the Attic year begins at the summer solstice autumn is early and spring late.

vine and abundance in the fruit.

This, then, is pruning of a universal sort and applies to all vines.

Pruning: The Seasons

As for the season of dressing the vines, (1) some distinguish them by the countries alone, (2) others[1] touch on the varieties. 15.1

(1) We are told to dress vines early[2] in dry and hot country, as soon as they are done with shedding the leaves, but in cold and rainy country to dress the vines shortly before they sprout. For in dry country the vines thus keep their fluid[3] and become better themselves and their fruit yields pleasanter wine; whereas in cold country the vines themselves are none the worse for losing their water and the fruit is less apt to decompose and makes pleasanter wine. In intermediate and well-tempered country we should dress the vines at a point intermediate between the two seasons.

Some however recommend this early[4] dressing for all vines without distinction (a point mentioned before)[5] because (1) the branch gets thicker[6] and 15.2

---

[3] By not bleeding (as a result of pruning) until the fruit has been produced.
[4] In autumn.   [5] *CP* 3 13. 2.
[6] Because in winter the growth does not pass to leaves or fruit.

THEOPHRASTUS

ἀμβλῶπες[1] οὐκ ἔσονται, καὶ ἐν εὐδίᾳ συμβήσεται τὰς τομὰς ἐπιξηραίνεσθαι, πάντα δὲ ταῦτα συμφέρει· χειμῶνος μὲν γὰρ οἱ πολλοὶ τῶν ὀφθαλμῶν ἀπόλλυνται ῥίγει, καὶ ἡ τομὴ βίᾳ ξηραινομένη ῥήγνυται, τοῦ δ' ἔαρος ῥεῖ τό τε δάκρυον, καὶ ἀμβλῶπες[2] ἐπιγίνονται πολλοί.

15.3 καίτοι τινὲς οὐ διὰ ταῦθ' ὑπολαμβάνουσιν, ἀλλ' ὡς τίκτουσαν τὴν ἄμπελον εὐθὺς ὅταν τρυγηθῇ, καὶ τότε γινόμενον τὸν καρπόν (οὐ γὰρ οἷόν θ' ἅμα δύο γόνους ἔχειν), οὐκ ὀρθῶς λέγοντες· γίνεσθαι γὰρ θέρους τὸν καρπόν, ὅταν ᾖ πολὺ τὸ ὑγρόν, ἐκ τούτου γὰρ τὴν γένεσιν εἶναι. τεκμήριον δέ· ἐὰν γάρ[3] τις ἐργάσηται τὰς τοῦ χειμῶνος ἀργὰς[4] ἀρχομένου τοῦ θέρους, εὐθὺ τὰ κλήματα πυκνόφθαλ-

---

[1] ego : ἀμβλωπὲς U N aP : ἀμβλωπεῖς u.
[2] ego : ἀμβλωπὲς U : ἀμβλωπεῖς u.
[3] Heinsius : δὲ U.
[4] Itali : ὄργας U.

---

[1] The word *amblṓps* ("that fail to bear properly") is probably connected with *amblýs* ("blunt," "dim" of vision). The metaphor was suggested by *ophthalmós* ("eye"), the word for bud. It is conceivable that the word *amblṓps* was also interpreted as connected with *amblóō*, "to miscarry."

[2] That is, weather that is not, like that of winter, wet and cold.

DE CAUSIS PLANTARUM III

the vine will thus have no shoots that fail to bear properly,[1] and because (2) the cut will dry in fair weather.[2] All of this is good: thus if the vine is dressed in winter most of the new buds perish of cold and the cut is forcibly dried[3] and cracks open; again when the vine is dressed in spring the cuts bleed and many buds are put out that fail to bear properly.

Some persons nevertheless do not take these to be the reasons for dressing the vine early, but suppose that the reason is that the vine is in the process of parturition directly it is harvested and the fruit is then in the process of being produced (since it is impossible, they urge, for the vine to have two broods at the same time).[4] But they are mistaken (it is argued) since the fruit is generated in summer, when the fluid in the plant is abundant,[5] for it is from the fluid that the fruit is produced, and here is proof: if at the beginning of summer you till vines that have been left untilled through the winter, the branches at once become dense with buds owing to

15.3

---

[3] By the cold; for its drying effect *cf. CP* 2 8. 1–2.

[4] That is, the new fruit cannot be produced until the old fruit is removed. The pruning therefore must be carried out in autumn, when the new fruit is on the way, so as to benefit it at once.

[5] The fluid has come from the rains of winter and spring and has not been dried out by the rainless summer.

μα¹ γίνεσθαι διὰ τὸ πλῆθος τοῦ ὑγροῦ, τοῦτο δὲ τὴν ἐργασίαν ποιεῖν ὅτι πληροῖ, τὸν δ' ἥλιον θερμαίνοντα πηγνύναι, καὶ πυκνοτέρους τοὺς ὀφθαλμοὺς γίνεσθαι καὶ βότρυς πλείους.

15.4 οἱ δέ, ταῦτα² κελεύοντες, ἄλλως πως τὰς αἰτίας λέγουσιν· φασὶ³ γὰρ τοῦτο δεῖν ἐν τῇ θερμῇ καὶ λεπτῇ, καὶ διὰ τοῦτο πρωὶ δρᾶν,⁴ ὅπως τὸ ὕδωρ εἰς τὰ καιριώτατα τῶν κλημάτων ἔλθῃ, καὶ ἔχῃ τροφὴν ἐν τῷ θέρει· τὰς δ' ἐν τοῖς ἐφύδροις, ἢ ὅσα τῶν γενῶν ὑβριστικά, τοῦ ἦρος, ὅπως διεσκεδασμένον ᾖ τὸ ὑγρόν,⁵ καὶ τμηθείσης ἐν ὥρᾳ, τοῦτ' ἀπορρυῇ· διὰ γὰρ τὸ πλῆθος οὐ πεττούσας τοῦθ',⁶ ὑβρίζειν ἀλλὰ⁷ καὶ ἐκκληματοῦσθαι.⁸ φυτοῖς δ' οὖσι τοῖς τοιούτοις καὶ ἐπισπείρειν ἐπὶ τὰ ἄνδηρα⁹ δεῖν¹⁰ κριθὰς καὶ κυάμους, ἐπειδὴ ταῦτα ξηραντικώτατα¹¹ πάντων.

¹ u : πυκνοφθαλμία U.
² U : ταὐτὰ Schneider.
³ aP : φησὶ U N.
⁴ ego (ποιεῖν Wimmer : καὶ Schneider) : καν U.
⁵ ego (διεσκεδασμένου τοῦ ὑγροῦ Schneider) : διεσκεδασμένη τὸ ὑγρὸν U (διεσκεδασμένη τὸ ὑγρὸν u N) : διεσκεδασμένον τὸ ὑγρόν aP.
⁶ U : τόθ' Heinsius.
⁷ ego (ἄλλως Wimmer) : ἀλλὰ U : ἀλλὰ u.
⁸ Schneider : ἐγκληματοῦσθαι U.
⁹ Schneider : ἀνθηρᾶ U.

the great amount of fluid in them, and for this the tillage is responsible, since it fills them with it; the sun on the other hand warms the fluid by its heat and makes it firm, and the buds acquire more consistency and the clusters become more numerous.

(2) Others make the recommendation[1] but give the reasons somewhat differently. They say that it is in hot and thin country that we must dress the vine early,[2] and do it early on this account: so that the rain[3] may get to the branches that matter most[4] and these may have it as food in summer[5]; but to dress vines in regions with surface water, or the varieties apt to luxuriate, in spring, so that the fluid will have become dispersed throughout the vine,[6] and the fluid so dispersed may bleed away with the vernal pruning, since it is owing to the great amount of this fluid that the vines fail to concoct it and get out of hand in various ways, among them running to branch. When such varieties are slips one should sow on the borders barley and beans, these having the greatest drying power of all plants.[7]

15.4

---

[1] *CP* 3 15. 2–3 (to prune in autumn). [2] In autumn.
[3] Of winter. [4] The rest are pruned away.
[5] The following summer, when the vine bears.
[6] Instead of remaining in the roots, as in winter.
[7] *CP* 2 18. 1.

---

[10] U^{ar} : δεῖ U^r N aP.   [11] U^{cc} (ώ from ὰ).

15.5 αἱ μὲν οὖν αἰτίαι σχεδὸν αὗται παρ' ἑκάστων.

ἀκριβεστάτη δὲ ἡ ἐν ἀμφοῖν διαίρεσις, ἔν τε ταῖς χώραις καὶ τοῖς γένεσι, καὶ πρὸς τὰς ὥρας καὶ πρὸς τὸ βραχυτομεῖν ἢ μακροτομεῖν · ἔνια γὰρ οὐ φέρουσιν ἂν[1] βραχυτομηθῶσιν,[2] ἀλλ' εἰς τὴν βλάστησιν τρέπονται, καθάπερ ἡ Ἀφυταῖος καλουμένη, καὶ ἐν Ἀκάνθῳ δὲ ἐπὶ τέτταρας ὀφθαλμοὺς ἐλαχίστους τέμνουσιν · αἱ δὲ μακροτομούμεναι ταχὺ καταγηράσκουσιν διὰ τὴν πολυκαρπίαν. νέας δ' οὖν οὔσας ἀναγκαῖον βραχυτομεῖν, ὅπως καὶ ῥιζωθῶσιν καὶ αὐξηθῶσιν.

καὶ περὶ μὲν ἀμπελουργίας ἐκ τούτων ἄν τις θεωρήσειεν.

16.1 ἐπεὶ δ' ὅμοιον τῷ τοιούτῳ, καὶ ὥσπερ δεύτερον, ἡ βλαστολογία, δεῖ καὶ[3] ταύτην εὐθὺ ποιεῖσθαι καὶ βλαστολογεῖν ὅταν διαφαίνωσιν τὸν καρπόν · μετὰ δὲ ταῦτα εὐθὺ τὸ δεύτερον σκάπτειν, ὅπως ὅ

---

[1] ego : φέρουσι καν U.
[2] U : βραχὺ τμηθῶσιν u.
[3] a : δεῖσθαι U N P.

---

[1] CP 3 14. 2–3 15. 4.

DE CAUSIS PLANTARUM III

So these (one may say) are the reasons given by the various schools.[1] 15.5

The most exact distinction is the twofold one of both country and kind, both for determining the seasons and for determining whether to prune short or long. For some kinds will not bear if pruned short, but run to vegetative growth, like the so-called vine of Aphytis[2]; so at Acanthus[3] too the growers always prune so as to leave at least four buds. On the other hand when the branches are pruned long the vine soon ages from abundant bearing. With young vines in any case we are compelled to prune short, so that the vines may both root and grow.[4]

These points, then, may serve for the study of vine-dressing.

The Remaining Operations

Since thinning the shoots[5] is similar and (as it were) a second dressing, this operation too must be carried out with no delay as soon as the vines show promise of fruit. Right after this we must spade[6] for the second time, to let both the fruit and the 16.1

---

[2] A city in the Chalcidice.
[3] Doubtless the city in the Chalcidice.
[4] *Cf. CP* 3 13. 1.
[5] *Blastologia* ("thinning"), literally "selecting shoots," is the removal of flower clusters at various stages before the formation of the grapes.
[6] For the first spading *cf. CP* 3 12. 2.

τε καρπὸς καὶ ὁ βλαστός, νέος ὢν καὶ ἐν ὁρμῇ τοῦ βλαστάνειν, ὡραίαν λάβῃ τὴν σκαπάνην· ἔπειτα πάλιν βλαστολογεῖν πρὸ τοῦ ἀνθεῖν, συμβαίνει γὰρ ἐν τούτῳ τὸ βοστρύχιον[1] αὔξεσθαι, διὰ τὸ μήπω συνεστάναι τὰς ῥᾶγας· ὅταν δὲ ἀπανθήσῃ, τὸ μὲν συνέστηκεν, αἱ δὲ συνίστανται καὶ αὔξονται.

16.2 τὸ δ' ὅλον ἀκμαῖα[2] μάλιστ' ἀποδίδοσθαι ζητεῖ τῶν ἔργων ἥ τε βλαστολογία καὶ ὁ σκαφητὸς ὁ μέσος, τῇ γὰρ ἀμπέλῳ τότε συμβαίνει πρὸς αὔξησιν ὁρμᾶν τῶν τε παρόντων καρπῶν, καὶ τῶν βλαστῶν ἐν οἷς ἄρχεται γονεύειν τὸν εἰς νέωτα καρπόν. ἐὰν οὖν ὁ καρπὸς[3] μὴ παρεθῇ, καλῶς γονεύσει. βλαστάνει δὲ μέχρι που[4] Κυνὸς ἐπιτολῆς, ὅταν δὲ ἐπιτείλῃ, παύεται τὰ μέλλοντα καλῶς ἐγκύμονα γίνεσθαι· θερμαίνοντος γὰρ τοῦ ἡλίου τὸ προηγούμενον τῶν βλαστῶν ἁπαλὸν ἀποσκληρύνεται καὶ παύεται· πεπαυμένης δὲ τῆς αὐξήσεως, φύσιν ἔχειν[5] συνίστασθαι τὴν ὑγρότητα

---

[1] Gaza (*pendula racemi*), Scaliger : στρύχιον U.

[2] Itali after Gaza (αὐκμαῖα u) : αὐχμαῖα U (-α miswritten N) aP.

[3] U : σκαφητὸς Gaza (*fodiendi cura*), Itali : καιρὸς Wimmer.

[4] ego : τοῦ U N : τῆς aP.

[5] ἔχειν U : ἔχει u : ἔχει τὸ Wimmer.

shoot,[1] when it is young and in the process of coming out, profit in time from the spading. Next we must thin again before the vine blooms, since it happens that at this time the "ringlet"[2] is in the process of growth, since the grapes are not yet formed, but after the flower is shed the formation of the ringlet is completed and the grapes are now in the process of formation and growth.

In general the measures that most require timely application are the thinning and the second of the three spadings,[3] since it happens that then the vine feels the impulse to grow both the fruit already present and the shoots in which it is beginning to generate the fruit of the following year. So if the fruit is attended to when it is still forming the vine will generate properly. It keeps adding to the growth of the shoots until about the rising of Sirius, but after that the parts that are to have a good pregnancy stop growing, since with the heat of the sun the delicate end of the shoot gets hardened and stops, and once the growth ceases the fluid of the shoot (it is said) tends naturally to become set, and

16.2

---

[1] That is, the shoot which will bear next year (*cf. CP* 3 16. 2).

[2] The inflorescence.

[3] For the first spading *cf. CP* 3 12. 2; for the second *cf. CP* 3 16. 1; for the third spading (the "dusting") *cf. CP* 3 16. 3.

THEOPHRASTUS

τοῦ βλαστοῦ, καὶ ἐκ ταύτης φύεσθαι τὸν καρπόν. ὅσα δ' ἐν παλινσκίοις ἢ ἐνύδροις[1] ἐστί, κρατεῖ τοῦ
16.3 ἡλίου καὶ πλείω χρόνον αὔξεται. καιρὸν δέ τινα ζητεῖ καὶ ἡ διαστολὴ καὶ ἡ κόλουσις, οὐ μὴν ἄσημόν γ' ὁμοίως οὐδὲ χαλεπὸν καταμαθεῖν.

τὴν δ' ὑποκόνισιν, τὸ μὲν ἐν τοῖς πρώτοις καιροῖς, ὅταν ἄρχωνται περκάζειν οἱ βότρυες, μὴ ἐᾶν ἄχρι οὗ πεπανθῶσιν, ὀρθῶς ἔχει· κωλύουσι γὰρ τὴν ὁρμὴν τῆς πέψεως, ἀντισπῶντες ἑτέρᾳ κινήσει (διὸ καὶ οὐδὲ τὴν πόαν οἴονται[2] δεῖν ἐκτίλλειν)· ὡς <δ'>[3] ὅλως ἀχρεῖον καὶ βλάπτον, οὐκ ὀρθῶς· ἥ τε γὰρ χρεία μαρτυρεῖ καὶ τὸ ἐφ'[4] ἑτέρων γινόμενον· οἱ γὰρ σίκυοι δοκοῦσι τρέφεσθαι τούτῳ καὶ ἁπαλώτεροι γίνεσθαι κατακρυπτόμενοι τῷ κονιορτῷ, διὸ καὶ Μεγαρεῖς[5] κρύπτουσι.

16.4 θαυμάζεται δ' εἰ ξηρὸς ὢν τρέφει· τὸ δ' αἴτιον

---

[1] U : ἐφύδροις Schneider.
[2] u : οἰόντε U : οἰόν τι N aP.
[3] Heinsius after Gaza.
[4] N aP : ὑφ' U.
[5] u aP : μεγαρεῖς U : μεγάρης N.

---

[1] The same as training, for which see *CP* 3 13. 1–4.
[2] The same as pruning: cf. *CP* 3 14. 1–3 15. 5.
[3] The same as the third spading; for the three cf. *CP* 3 16. 2. Cf. *HP* 2 7. 5, cited in note 2 on *CP* 3 10. 1.

DE CAUSIS PLANTARUM III

from this fluid comes the fruit. Vines however in shady or well-watered land prevail over the sun and keep growing longer. Spreading the vine[1] and removal of parts[2] also require timing, but here there is no such absence of guidance and no such difficulty in determining the right moment.

16.3

As for dusting[3] the fruit, we get one good piece of advice and one bad. The good advice is not to neglect it at first, when the clusters are beginning to turn dark, and wait until they are ripe; for the notion behind the advice is to slow down the movement toward concoction by opposing to this movement another (which is why the experts would not even have the weeds pulled up).[4] The bad advice is to dismiss dusting as quite useless and in fact harmful,[5] since its usefulness is attested by growers' resorting to it and by results in other plants: so gourds are considered to be fed by dust and become tenderer when covered with it, which is why the Megarians hoe dust over them.[6]

That dust, which is dry, should feed a plant, is

16.4

---

[4] Presumably the delay in concoction allows more fluid to collect in the fruit.

[5] *Cf. HP* 2 7. 5: "But some deny that one should dust the vine or touch it at all when the cluster is turning dark, but only (if at all) when it has become dark; and some hold that one should not even touch the vine then, except to pull out the weeds."

[6] *Cf. HP* 2 7. 5, cited in note 2 on *CP* 3 10. 1.

ἴσως ἐκ πλειόνων ἂν εἴη· καὶ τῷ κινουμένης τῆς γῆς ἀναδίδοσθαι τροφὴν ἐνίκμου τινὸς ἀέρος (ὥσπερ ἐπὶ τῶν αὐχμωδῶν[1] ἐλέχθη), καὶ τῷ προβολὴν ἅμα πρὸς τὸν ἥλιον εἶναι· καὶ ἔτι καταξηραινόμενος[2] ἁπαλῇ ξηρότητι καὶ μαλακῇ μᾶλλον ἐπισπᾶται τὴν τροφὴν ἐκ τῆς ἀμπέλου· πάντα γὰρ ταῦτα συνεργεῖ καὶ πρὸς εὐτροφίαν καὶ πέψιν.

ἡ δὲ τῶν σικύων κατάκρυψις[3] οὐκ ἄλογος· ἀναξηραίνων γὰρ ὁ ἥλιος σκληρύνει (διὸ καὶ ὑπὸ τὰ φύλλα κρύπτουσιν)· ὥστ', ἐπιβολὴν ἔχοντες καὶ βρεχόντων,[4] ἀμφοτέρως[5] εὐτραφεῖς[6] γίνονται καὶ ἁπαλοί.

καὶ περὶ μὲν ἀμπέλων ἱκανῶς εἰρήσθω.

17.1 πάντων δὲ ἰδιώτατον εἶναι δοκεῖ τῶν[7] κατὰ τὰς θεραπείας οἱ ἅλες οἱ τοῖς φοίνιξι παραβαλλόμενοι, καὶ γὰρ πρὸς εὐβλαστίαν καὶ πρὸς εὐκαρ-

---

[1] Schneider : ἀτμοδῶν U.
[2] Schneider : εἴ τι καταξηραινόμενον U.
[3] u : κατάκυψις U.
[4] ego (βρεχόμενοι Schneider) : βρεχομένων U.
[5] Schneider : -ων U.
[6] u aP (εὐτραφαῖς N) : εὐτροφεῖς U.
[7] u : τὸν U.

---

[1] CP 3 10. 1.

DE CAUSIS PLANTARUM III

wondered at. The explanation would perhaps come from several circumstances: not only is food, consisting of a kind of air with moisture in it, given to the plant when the soil is turned up (as we said[1] in speaking of rainless country), and not only is dust at the same time a screen against the sun, but furthermore the cluster, when dried by a delicate and gentle dryness, is better at drawing the food from the vine, since all these circumstances work together for both good feeding and concoction.

The covering of cucumbers with dust is not unreasonable. For the sun dries them out and so makes them hard, which is why growers also hide them under the leaves. So there are two ways for them to become plump and tender: you cover them or water them.[2]

Let this discussion suffice for the vine.

Salting the Date-Palm[3]

Of all agricultural procedures the most unique is held to be the application of lumps of salt to the date-palm, since the procedure promotes not only

17.1

---

[2] *Cf. HP* 2 7. 5, cited in note 2 on *CP* 3 10. 1.
[3] *Cf. CP* 2 5. 3 and *HP* 2 6. 2 (of the date-palm): "... it likes briny ground. Hence where the ground is not briny the growers sprinkle salt round the tree. One must not do this close to the roots, but at a distance, and sprinkle about a twelfth of a medimnus."

THEOPHRASTUS

πίαν χρήσιμον[1] (ἐπεὶ ὅσα γε πρὸς ἰατρείαν τινὰ τῶν τοιούτων ἢ φυλακήν, οὐδὲν ἄτοπον, οἷον ἡ τέφρα ταῖς τε συκαῖς καὶ τῷ πηγάνῳ, ξηρὰ γάρ,[2] πρὸς τὸ μὴ σκωληκοῦσθαι, μηδὲ σήπεσθαι, τὰς ῥίζας · βοηθεῖ γὰρ καταξηραίνουσα).

17.2 περὶ δὲ τῶν ἁλῶν,[3] ἢ τῆς ἅλμης ἣν παρέχουσιν (οἱ μὲν γὰρ οὕτως ποιοῦσιν, οἱ δ' ἐκείνως [ποιοῦσιν][4]), ἁπλῆ μέν τις αἰτία, διότι φιλεῖ χωρία ἁλμυρώδη. σημεῖον δέ · ὅτι παρ' οἷς πλῆθος φοινίκων, ἐν τούτοις ἐστίν, οἷον Λιβύῃ Συρίᾳ ταῖς ἄλλαις, ὥστε καθάπερ[5] οἰκείαν τινὰ βοήθειαν βοηθοῦσι διὰ τῶν ἁλῶν.[6] ὅτι μὲν γὰρ ὁ φοῖνιξ χαίρει τῷ ἁλμώδει, κοινή τις <ἂν> αἰτία λέγοιτο[7] περὶ πάντων ὅσα ζητεῖ τόπους διαφόρους, ἀπὸ τῶν πρώτων καὶ μεγίστων ἀρχομένοις, οἷον ἐνύδρων χερσαίων, καὶ ὅσαι δὴ καθ' ἑκάτερα τούτων δια-
17.3 φοραί · δῆλον γὰρ ὡς τῇ κράσει πως σύμμετροί τε

---

[1] N aP : χρήσιμοι U.   [2] γὰρ U : Schneider deletes.
[3] u : ἄλλων U N aP.   [4] Schneider.
[5] Wimmer : ὥσπερ καθάπερ U N : καθάπερ γὰρ aP.
[6] u N P^c : ἄλλων (ἀλλων U) aP^{ac} (?).
[7] Wimmer : αἰτία λέγοιτο u (no accents U) N : αἰτία λέγοιτ' ἂν aP.

---

[1] *Cf. HP* 2 7. 6 (treatment of running to leaf): "In the fig in addition to cutting off roots they scatter ashes round it . . ."

DE CAUSIS PLANTARUM III

foliage but fruitfulness. As for such measures as have a curative or preventive effect, there is nothing strange about them, for instance the application of ashes (since they are dry) to fig-trees[1] and rue[2] to keep the roots from worms and decomposition, the ashes acting against this by drying the roots.

As for the salt or brine that they apply—some do it with salt, some with brine—there is a simple reason: the date-palm likes salty ground. Proof is the presence of salty ground wherever date-palms abound, as in Libya, Syria and elsewhere[3]; and so when the lumps of salt are applied the help is (as it were) from home. As for the reason why the date-palm likes salinity, a general sort of reason could be given, one of common application to all plants that seek special localities, beginning with the first and largest classes, aquatic and terrestrial,[4] and the various subdivisions of each: the regions evidently

17.2

17.3

---

[2] *Cf. CP* 5 6. 10.

[3] *Cf. HP* 2 6. 2 (of the fondness of the date-palm for salty ground): "They also take as proof of its liking such ground the fact that wherever date-palms abound the ground is salt; so they say that in Babylonia it is salty where the date-palms grow, and also in Libya and Egypt and Phoenicia, and in the valley of Syria where their numbers are greatest the dates that can be kept are produced in only three districts, and these are salty ..."; *cf.* also *HP* 4 2. 5 for the association of the tree with brine in Libya.

[4] For the distinction see note *a* on *CP* 2 3. 5.

THEOPHRASTUS

καὶ οἰκεῖοι καθ' ἕκαστα τυγχάνουσιν, ὥσπερ ἐπὶ τῶν ζῴων· ὁ γὰρ αὐτὸς λόγος καὶ περὶ τούτων καὶ ἡ αὐτὴ ζήτησις.

οὐ μὴν ἀλλ' εἴ γέ τι καὶ περὶ τούτων ἴδιον χρὴ εἰπεῖν, ἐν ἀμφοῖν τὴν αἰτίαν ζητητέον, ἔν τε τῷ τὴν γῆν ποιάν τινα ποιεῖν, καὶ ἐν τῷ τὰς ῥίζας, ἅπερ ἄμφω συμβαίνει διὰ τῶν ἁλῶν· τὴν μέν γὰρ ποιοῦσιν κούφην καὶ χαύνην, τὰς δὲ ῥίζας εὐτραφεστέρας καὶ παχυτέρας, ὁλκοτέρας τε[1] ποιοῦντες τῷ πόρους τινὰς ἀνοίγειν καὶ ἔτι τῷ[2] καταψύχειν, ὅπερ ἡ κόπρος οὐ ποιεῖ διὰ τὸ ἔμπυρον.

17.4 ὁ δὲ φοῖνιξ τὸν μὲν ἀέρα τὸν περιέχοντα ζητεῖ θερμόν, οὕτως γὰρ ἡ πέπανσις τοῦ καρποῦ, τὰς δὲ ῥίζας καταψύχεσθαι, διὰ τὴν ξηρότητα, τοῦτο δὲ ποιοῦσιν οἱ ἅλες. ὅθεν καὶ τοῖς μὲν χρῶνται Βαβυλώνιοι, τῇ κόπρῳ δὲ οὐ[3] χρῶνται.

τοῦ[4] δὲ μόνῳ τούτῳ συμφέρειν τὴν ἰδιότητα τῆς φύσεως αἰτιατέον· ὥσπερ γὰρ τὸ ξύλον, καὶ

---

[1] U : ἑλκτικωτέρας Wimmer. [2] τῷ u : τὸ U.
[3] U^{css} : U^{t} omits. [4] U^{cc} : το U^{ac} : τῷ N a(τῷ)P.

---

[1] *Cf. HP* 2 6. 3 (of the date-palm): "... as to manure there is a dispute. Some say that it does not like manure and that manure is instead extremely bad for it; others say that they employ it ..., but that one must give it plenty of water poured over the manure, as is done in Rhodes."

DE CAUSIS PLANTARUM III

by their tempering of qualities somehow happen to be just right and appropriate for the various different plants, just as they are for the different animals (for the same explanation and the same question applies to animals too).

Still if we must give some special explanation for this phenomenon as well, we must look for the reason in both things: in the imparting of a certain quality to the earth and again to the roots, and both results are effected by the salt, for it makes the earth light and loose, and the roots plumper and thicker, since it not only renders them more capable of attracting food by opening certain passages in them, but again cools them (which manure does not do because of its great heat).[1] And although the date-palm likes heat in the surrounding air, since the ripening of the fruit depends on this, it likes its roots cool because of its dryness, and this is done by the salt. Hence the Babylonians use salt but no manure.[2]

17.4

That salt is good for this tree alone must be accounted for by its special nature; for just as its wood differs from all other wood,[3] so too its roots

---

[2] Manure is heating: *cf. CP* 3 6. 1.

[3] Through its own nature it gives an evil smoke (*HP* 5 9. 4). It was supposed to arch upward under pressure: *cf. HP* 5 6. 1; Aristotle, Fragment 229 (ed. Rose³); Plutarch, *Natural Questions*, chap. xxxii; Xenophon, *Cyropaedia*, vii. 5 11. The notion may have owed something to the Greek word for "resist" (ἀντωθεῖν), literally "push back."

THEOPHRASTUS

αἱ ῥίζαι διάφοροι τῶν ἄλλων, διόπερ ἀναδηχθεῖσαι[1] μᾶλλον ἐπισπῶνται τὴν τροφήν, καὶ αὐταὶ[2] ἀρκοῦνται[3] καὶ τὸ ὅλον δένδρον αὔξουσιν.

17.5 ποιεῖ δὲ καὶ ὡς[4] ἡ γεωργία τὰς ῥίζας πάντων ὁλκοτέρας[5] (οὕτω γὰρ καὶ ἐξημεροῦνται τῶν ἀγρίων ἔνια), καὶ ἡ κόπρος δ' <ἡ>[6] ἰσχυροτάτη τὸ αὐτὸ τοῦτο δρᾷ, καθάπερ ἡ σκυτοδεψικὴ τὰς τῶν μυρρίνων καὶ οὖρον παραχεόμενον, ἄμφω γὰρ διαδύνεται μᾶλλον (αἱ δὲ ῥίζαι δέονται τῆς ἀναδήξεως καὶ τῶν ἄλλων), ὡσαύτως καὶ τῶν ῥοιῶν· καὶ γὰρ ταύταις παραχέουσι καὶ τὴν σκυτοδεψικὴν παραβάλλουσιν, πλὴν οὐχ ὁμοίως καὶ τοῖς
17.6 μυρρίνοις, ἀλλ' ἧττον. αἱ δ' ὑγραὶ καὶ σαρκώδεις οὐ δέονται τῆς ἀναστομώσεως καὶ δήξεως.

ὅλως δ' αἱ μεταβολαὶ τῶν δένδρων διὰ τὰς θεραπείας ὥστε ἐξ ὀξέων καὶ πικρῶν γλυκέα

§ 5 line 7 *Geoponica*, iii. 3. 4: τούτῳ τῷ μηνὶ (March) ταῖς ῥίζαις τῶν ἀμυγδαλῶν κόπρον χοιρείαν ἐπιθήσομεν· τάς τε γὰρ πικρὰς γλυκείας ποιεῖ καὶ μείζους καὶ τρυφεράς, ὡς Ἀριστοτέλης (Frag. 277 Rose[3]) φησί. Θεόφραστος δὲ οὖρον ταῖς ῥίζαις ἐπιχέειν φησί.

[1] u : ἀναδειχθῆσαι U : ἀναδειχθεῖσα N : ἀναδειχθεῖσαι aP.
[2] Scaliger (*sibi* Gaza) : αὗται U^c from αὔ-.
[3] U : αὔξονται Link.
[4] U : Gaza omits : ὡς <οἱ ἅλες> Schneider.
[5] Schneider : ἀλκιμωτέρας U.

differ from all others, which is why they attract food better on being irritated, and so get their need of it themselves and make the whole tree grow.

Further Cases of Improvement Through Irritation of the Roots

Just as husbandry makes the roots of all plants better able to attract (this being how some wild plants[1] are turned into cultivated ones), so again the strongest manure does the same, as when tanner's manure and urine are applied to the roots of the myrtle,[2] since both substances are more penetrating (but the roots of the rest require irritation too[3]) and so also do those of the pomegranate,[4] for here too urine is poured and tanner's manure applied, although not to the same extent as with the myrtle, but less. Succulent and fleshy roots on the other hand do not require this opening of passages and irritation.

17.5

17.6

Indeed the changes by husbandry that turn a tree from sharp and bitter to sweet,[5] as with pome-

---

[1] Cf. HP 2 2. 9, 2 2. 12; CP 2 14. 2. At HP 3 16. 1–2 filbert is mentioned (which is transplanted); at HP 2 4. 1 wild barley and wild wheat.   [2] Cf. CP 3 9. 3.

[3] At CP 3 9. 2 almond and olive are mentioned in this connexion (besides myrtle and pomegranate).

[4] Cf. CP 3 9. 3; 2 14. 2.   [5] Cf. CP 3 9. 3.

---

[6] Schneider : δ' U (δι̇ N : δὲ P : δὴ a).

THEOPHRASTUS

γίγνεσθαι, καθάπερ αἵ τε ῥόαι καὶ αἱ ἀμυγδαλαῖ καὶ εἴ τι ἄλλο, τῷ[1] τὰς ῥίζας ἀλλοιοῦσθαι · μεταβάλλουσιν οὕτω τὴν γῆν καὶ τὴν τροφήν.[2] ταύτῃ μὲν γὰρ[3] ἐπὶ[4] τρία ἢ τέτταρα ἔτη γίνονται,[5] ἐπὶ τοσοῦτον γὰρ ἡ θεραπεία · ῥόᾳ δ' αἱ ῥίζαι[6] δια-
17.7 μένουσιν καὶ τὸν[7] πλείω χρόνον. μεταβολὴ δ' εὐλόγως τῶν καρπῶν ἐκπεττομένης μᾶλλον τῆς τροφῆς · ἐκπέττεται δὲ τῷ τοιούτῳ,[8] <τῷ>[9] ποιάν τινα τὴν ῥίζαν εἶναι. τὸ δ' ὅλον (ὥσπερ καὶ πρότερον ἐλέχθη περὶ τῶν σπερμάτων) κινουμένης καὶ μεταβαλλούσης τῆς ἀρχῆς συμμεταβάλλει καὶ τἆλλα, καὶ ἡ ὅλη φύσις · αἱ τροφαὶ δὲ κινοῦσι καὶ ἐξαλλοιοῦσιν ἐὰν πλείω γίνωνται χρόνον. (ἐπεὶ [δὲ][10] καὶ ὅσα τοῖς θερμοῖς ἀρδευόμενα βελτίω, καθάπερ ἥ τε μηλέα δοκεῖ ἡ ἠρινὴ καὶ ὁ μύρρινος · καὶ γὰρ οὕτως[11] ἀπύρηνος, ὥς φασιν,

---

[1] τῷ u : το U (τὸ N aP) : <ὥσπερ> τῷ Schneider.
[2] μεταβάλλουσιν ... τροφὴν U : οὕτω καὶ τὴν γῆν καὶ τὴν τροφὴν μεταβάλλει Schneider : μεταβάλλει · μεταβάλλουσι δὲ κατὰ τὴν γῆν καὶ τὴν τροφὴν Wimmer.
[3] ego (γλυκέα μὲν οὖν Wimmer) : αὕτη μὲν γὰρ U.
[4] U : μετὰ Schneider.
[5] U N : γίνεται aP.
[6] ego (radices punicae Gaza : ῥόας δὲ ῥίζαις Schneider : ῥόας δὲ πικραὶ Wimmer) : ῥοαι δε ῥίζαι U.
[7] τινα Wimmer : Schneider omits.

granate and almond and any other there may be,[1] are due to alteration in the quality of the roots[2]; and it is to bring this about that growers make changes in the soil and in the food. For the almond these changes last from three to four years (for the special tendance is kept up that long), whereas in the pomegranate the roots remain unaltered even longer. It is reasonable that the fruit should change with more thorough concoction of the food; and the better concoction comes from this: that the root has a certain quality. And in general (as was said earlier[3] about seeds), when the beginning is influenced and changes, the rest of the plant and its entire nature changes along with it; and the type of food influences the beginning and alters its quality if maintained for some time. (Indeed there are trees that improve on being watered with warm water, as the spring apple[4] and myrtle are considered to do. In fact under this treatment too[5] the myrtle, they

17.7

---

[1] No other is elsewhere mentioned: *cf. HP* 2 2. 9, 11 (reading at 2 2. 9 καθάπερ ἥ τε for Wimmer's ἀπαγριοῦται and καὶ ἀπορῇ τε of U). [2] *Cf. CP* 2 14. 2–3.

[3] *CP* 2 14. 3; 3 9. 4.

[4] This watering with warm water is not mentioned elsewhere in the *CP* or *HP*. Conceivably the warm springs of Thrace (*CP* 2 5. 1) may be meant.

[5] Manuring does the same: *cf. CP* 3 9. 3, 3 17. 5.

---

[8] [τοιοῦτῳ] Gaza, Schneider.   [9] ego.
[10] Gaza, Schneider.   [11] U N : οὗτος aP.

THEOPHRASTUS

γίγνεται, καὶ ἐξ ἀρχῆς οὕτω συνώφθη κατὰ σύμπτωμα, πρὸς λουτρὸν[1] ὄντος μυρρίνου καὶ ἐξημελημένου · τούτου γὰρ ἀπυρήνου γενομένου, λαμβάνοντες ἐφύτευον, ὅθεν τὸ γένος Ἀθήνησι γέγονεν.)

ἀλλὰ ταῦτα μὲν ἔχοντά τινα ὁμοιότητα πρὸς πίστιν εἴρηται τῶν ἐξ ἀρχῆς.

17.8 ἡ δ' ἁλυκότης τῶν ὑδάτων ὅτι καὶ τῶν λαχανωδῶν [τινα][2] τισιν ἁρμόττει, καὶ διότι τῷ λίτρῳ χρῶνται πρὸς ἔνια, πρότερον εἴρηται, διὸ καὶ τοῦθ', ὡς ἔοικεν, ὡς[3] οἰκεῖον παραληπτέον · ἀπὸ γὰρ τούτων δῆλον ὅτι καὶ τῆς τροφῆς καὶ τούτων ἡ γλυκύτης. ἀλλὰ τοῖς μέν, ᾗ[4] πικρότης τις ἐν τοῖς χυλοῖς σύμφυτος, τῷ δὲ φοίνικι, στρυφνότης, ἣν ἐξαιροῦσιν εὐθὺς οἱ ἅλες ἐκ τῆς πρώτης ἀρχῆς.

περὶ μὲν οὖν τῶν ἁλῶν[5] ἱκανῶς εἰρήσθω.

18.1 τὸ δὲ μὴ ἐπιμένειν ἐπὶ τῷ θήλει φοίνικι τὸν

---

[1] ego : λουτρῶν U^{ar} : -ῶι U^r (-ῶ N aP). [2] Itali.
[3] ego (Gaza omits : Schneider deletes) : ἐοικὸς U.
[4] ego (Schneider deletes) : η U : ἡ u. [5] u : ἄλλων U.

---

[1] It was not manured.
[2] *CP* 3 17. 5–7. [3] *CP* 3 17. 3–4.

DE CAUSIS PLANTARUM III

say, comes to have no stones. The idea of so treating it came by accident, from a myrtle facing a bath and neglected by the grower.[1] For when this myrtle turned out to have no stones, slips were taken from it and planted, and from these the variety at Athens has its origin.)

But all this,[2] bearing as it does a certain similarity to the case of the date-palm, is adduced to make the initial explanation[3] more convincing.

We said earlier[4] that salinity is also suited to some vegetables, and that soda is used with others. And so it seems we must accept the salinity here too as appropriate to the plants, since it is evident that the sweetness of these vegetables too comes from the saline water and the food. But the sweetness comes about in the vegetables insofar as their juices have a certain native bitterness[5]; and in the date-palm there is a certain astringency, which the lumps remove at the start from the beginning[6] of the tree.

17.8

Let this suffice for the discussion of the salt.

Another Unique Feature in the Care of the Date-Palm

That the fruit does not remain on the female

18.1

---

[4] *CP* 2 5. 3.
[5] *Cf. CP* 2 5. 4; 6 10. 8.
[6] That is, the roots.

THEOPHRASTUS

καρπόν, ἂν μὴ τὸ τοῦ ἄρρενος ἄνθος κατασείσωσιν ἅμα τῷ κονιορτῷ κατ' αὐτοῦ (καὶ γὰρ τοῦτο λέγουσί τινες), ἴδιον μὲν παρὰ τὰ ἄλλα, παρόμοιον δὲ τῷ ἐρινασμῷ τῶν συκῶν. ἐξ ὧν καὶ τὸ τελειογονεῖν μηδ' αὔταρκες εἶναι τὸ θῆλυ μάλιστα ἄν τις ἐπαγάγοι, πλὴν ἐχρῆν τοῦτο μὴ ἐφ' ἑνὸς ἢ δυοῖν, ἀλλ' ἐπὶ πάντων ἢ τῶν πλείστων εἶναι· τὴν γὰρ φύσιν οὕτω κρίνομεν[1] τοῦ γένους. ἀτοπώτατον δὲ καὶ ἐπ' αὐτῶν τούτων <τὸ> τοῦ[2] φοίνικος, ὁ γὰρ ἐρινασμὸς δοκεῖ φανερὰν ἔχειν τὴν αἰτίαν.

ἀλλὰ περὶ μὲν τούτων ἅλις.

18.2 ἡ δὲ τῶν ἄλλων ἰδιότης, ὅσων[3] ἐστίν, συμφανεστέρας ἔχει τὰς αἰτίας, οἷον ὡς τὴν ἀμυγδαλῆν,

[1] U^r N aP : κρίνωμεν U^{ar}.
[2] a : τοῦ U N : τὸ P.
[3] Schneider : ὅσον U.

---

[1] *Cf. HP* 2 6. 6 (of the date-palm): "... there are males and females; these differ in that the male first bears a flower on its spathe, whereas the female bears a small fruit directly."

[2] *Cf. HP* 2 8. 4 (of the remedy for fruit drop in the date-

date-palm[1] unless you shake the flower of the male over it together with the dust[2] (this too[3] being reported by some) occurs only in the date-palm, but is similar to the caprification of fig-trees.[4] From these instances one would be most inclined to infer that even a female tree cannot by itself bear completely formed fruit; except that this should hold not of just one or two female trees but of all or most of them, since this is how we decide the nature of the class of females. And in the cases before us that of the date-palm is very strange indeed, since caprification is considered to have a clear explanation.[5]

But enough of this.

Special Feature in the Care of Other Trees

Where there are distinctive features in the cultivation of other trees the reasons are more obvious.

18.2

---

palm): "It is carried out as follows: as soon as the male tree blossoms they cut off the spathe on which the flower is with no more ado and shake the downiness and flower and dust over the fruit of the female, and if this is done the female keeps its fruit and does not drop it. It appears that in both date-palm and fig a remedy comes to the females from the male (for they call the fruit-bearer the female), but in the date-palm the remedy is a sort of sexual union, in the other it works in a different way."

[3] The first report told of the application of salt to the roots (*CP* 3 17. 1).

[4] *Cf. CP* 2 9. 15 and *HP* 2 8. 4 (cited in note 2 above).

[5] *Cf. CP* 2 9. 5–6.

THEOPHRASTUS

ὅταν ἄρχηται καρποφορεῖν, οὔτε ὑδρεύουσιν οὔτε κοπρίζουσιν οὔτε διακαθαίρουσιν (εἰ μὴ τὰ αὖα μόνον), οὐδ' ἄλλο τῶν τοιούτων οὐδὲν δρῶσιν, πάντα γὰρ ἀφαιροῦσιν, ὅπως μὴ ὑπερισχύσασα ἀποβάλλῃ τοὺς καρποὺς ἐν τῷ ἄνθει,[1] διὸ καὶ χώραν οὐκ ἀγαθὴν ζητοῦσιν. ἐὰν δὲ μετὰ ταῦτα ἀκαρπῇ, τάς τε ῥίζας[2] γυμνώσαντες παραδιδόασιν τῷ χειμῶνι, καὶ τὰς ἄλλας κολάσεις προσφέρουσι τὰς εἰρημένας.

18.3 ὁμοίως δὲ καὶ ὅσων[3] οἱ φλοιοὶ περιαιροῦνται, καθάπερ ἀμπέλου, κεράσου,[4] φιλύρας, καὶ ὅλως τῶν φλοιορραγῶν · αὐτὴ[5] γὰρ ἡ φύσις ἐνταῦθα ἔοικεν ἐπιδεικνύναι τὸ συμφέρον, ἀφισταμένη καὶ ἀλλότριον ποιοῦσα · τὸ δ' ἀλλότριον ἅπαν βλαβερόν, ὥσπερ καὶ τὸ αὖον. καὶ τὸ ἐπὶ τῶν ἁλῶν[6] δὲ

---

[1] u : ανθη U : ἀνθεῖν Wimmer.
[2] τάς τε ῥίζας N.ac aP (N.cm inserts before this τάς τε ῥίζας ζητοῦσιν. ἐὰν δὲ μετα ταῦτα ἄκαρπας) : τας τὲ ῥιζας ζητοῦσιν. ἐὰν δὲ μετὰ ταῦτα ἄκαρπα τάς τε ῥίζας U.
[3] u : -ον U N aP.
[4] Schneider (from *HP* 4 15. 1) : κέδρου U.
[5] u aP : αὕτη U : αὐτῆ N.
[6] u : ἄλλων U.

---

[1] Such as spading and pulling out the weeds.
[2] *Cf. CP* 2 16. 8; 3 6. 8.
[3] *CP* 1 17. 9 with note *a* and 1 17. 10 with note *c*; 2 14. 1;

So when the almond starts to bear it is neither watered nor manured nor pruned (except only for removing the deadwood) nor subjected to any other such procedure [1]; for all such tendance is withdrawn so that the tree may not get too strong and drop its fruit in the blossom (this moreover is why the land sought for it is not of the best). [2] If after this it refuses to produce fruit they lay the roots bare for the winter to have its way with them and administer the other "castigations" that we have mentioned. [3]

So too with the trees whose bark is stripped off, as the vine, bird cherry, linden and in general trees that have bark which splits. [4] For the tree's own nature here appears to point out a good measure by divesting itself of the bark and making it an alien thing; and everything alien, like deadwood too, is harmful. Similarly with the lumps of salt: wherever

18.3

2 14. 4; 2 15. 4.

[4] *Cf. HP* 4 15. 1–2: "All trees die when the bark is removed all round ... except perhaps cork-oak ... Indeed the bark is also stripped from bird cherry, vine, and linden (from this ropes are made) ..., but it is not the important or inmost bark but the surface bark, which also sometimes falls off of its own accord because the other bark grows under it. Indeed some trees have bark that splits, as andrachne and plane. But, as some think, new bark grows under to replace this, whereas the outer bark dries out and splits and falls off of its own accord in many, but not so noticeably."

THEOPHRASTUS

τὸν αὐτὸν τρόπον · ὅπου τις ἰδιότης, οἰκεία πρὸς τὴν φύσιν.

καὶ περὶ μὲν δένδρων ἐκ τούτων θεωρητέον.

19.1 ἔχει δὲ ὁμοίως καὶ ἐπὶ τῶν στεφανωματικῶν, καὶ ὅλως τῶν φρυγανικῶν, σχεδὸν δὲ καὶ τῶν λαχανικῶν.[1] τὰ μὲν γάρ ἐστιν κοινὰ πᾶσιν, οἷον ὕδρευσις κόπρωσις[2] ἡ τοῦ ἐδάφους[3] ἐργασία, τὰ δὲ ἴδια καθ' ἕκαστον γένος. οἷον ἡ διακάθαρσις τοῖς φρυγανικοῖς ὁμοίως, ἐν στεφανώμασιν καὶ λαχάνοις, καὶ γὰρ ὁ ἕρπυλλος αὐαίνεται καὶ μὴ διακαθαιρόμενος καὶ τὸ σισύμβριον καὶ τἆλλα· τὰς δὲ δὴ ῥοδωνιὰς[4] οὐ κατακόπτουσι μόνον, ἀλλὰ καὶ ἐπικάουσιν, ἀνανθεῖς γὰρ γίνονται καὶ ἐκλοχμοῦνται, μὴ τοῦτο πάσχουσαι, δι' εὐτροφίαν

19.2 καὶ ἡ ῥάφανος δὲ καὶ τὸ πήγανον ἀποσκληρύνον-

---

[1] U : λαχανηρῶν Schneider.
[2] u (cf. HP 2 7. 1 : κόπρισις Scaliger) : κόπρωσις U N aP.
[3] Wimmer : ὕδατος U.
[4] ego : -ίας U.

---

[1] CP 3 2. 1–3 18. 3.
[2] For undershrubs cf. CP 3 7. 11. Cf. also HP 2 7. 1 of trees: "In the matter of cultivation and care some procedures are common, others are confined to this or that

DE CAUSIS PLANTARUM III

there is something special to the tree, it is appropriate to the tree's nature.

The measures taken with trees, then, are to be studied in the light of the foregoing discussion.[1]

Measures Taken with Undershrubs and Vegetables

The case is similar with coronaries, and indeed with undershrubs in general, and even (one might add) with plants of the vegetable kind. So some procedures apply to all, such as watering, manuring and working the soil, whereas others are special to a class. For example pruning the stems is found in coronaries and vegetables, just as it is in undershrubs,[2] for tufted thyme will wither when it is not also pruned, as will bergamot mint and the rest.[3] As for rose bushes, they are not merely pruned to the ground but also burned over,[4] since if this is not done they fail to flower and turn into thickets because of their rich feeding. Again cab-

19.1

19.2

---

tree. Spading, watering and manuring are common, and furthermore pruning the stems and removing deadwood."

[3] That is, the other coronaries which (like tufted thyme, bergamot mint and rose) are also undershrubs.

[4] *Cf. HP* 6 6. 6 (of the rose bush): "When burned over and cut short it bears a better flower, for when not disturbed it grows out and turns into a thicket."

THEOPHRASTUS

ται καὶ ἀποξηραίνονται, κολουσθέντα δὲ καὶ παλιμβλαστῆ γενόμενα, μείζω καὶ καλλίω καὶ εὐχυλότερα· δεῖ γὰρ καὶ τῶν χυλῶν καὶ τῶν ὀσμῶν ἀφαιρεῖσθαι τὸ ἄγαν δριμύ, τῇ γὰρ μεσότητι τὸ ἡδὺ καὶ τὸ σύμμετρον. ἐν δὲ τοῖς λαχάνοις ἀφαίρεσις μὲν οὐκ ἔστιν, πλὴν ὅσα κειρόμενα καλλίω, 19.3 καθάπερ τὸ πράσον ἢ εἴ τι[1] τῶν πολυβλαστῶν. ἡ δὲ τῶν παραφυομένων[2] ἐξαίρεσις, καὶ ἡ αὐτῶν τῶν ὁμογενῶν, ὅταν ᾖ πυκνά, καθαπερεὶ[3] διακάθαρσις γίνεται, τροφὴν πλείω ποιοῦσα καὶ αὔξησιν τοῖς λοιποῖς.

ἔχει δὲ καὶ ἰδιότητας ἔνια, καθάπερ ἐπὶ τοῦ πηγάνου πρότερον ἐλέχθη, καὶ ἐπ' ἄλλων ἐστίν.

ἀλλὰ ταῦτα μὲν ἐλάττω καὶ σχεδὸν φανερά.

20.1 λοιπὸν[4] δέ, καὶ ὥσπερ ἀντικείμενα τοῖς περὶ

---

[1] ἢ εἴ τι Wimmer (<καὶ> εἴ τι <ἄλλο> Schneider): ἤτοι U: ἢ τί u N (ἢ τὶ P): ἢ τι a.
[2] Schneider: παραφυτευομένων U.
[3] u: καθάπερ ἡ U N aP. [4] U: λοιπὰ Schneider.

---

[1] Cabbage is a vegetable; rue is a undershrub (*HP* 1 3. 2) and vegetable-like (*HP* 6 1. 2).

[2] This is not said elsewhere in the *HP* or *CP* of rue. For cabbage *cf. HP* 7 2. 4: "But this is admitted in the case of cabbage—that if it comes up again it is better eating ...";

bage and rue[1] get hard and dry, but when they have grown up again after being cut down they are larger, finer and juicier,[2] since the excess of pungency must be removed from juices as well as from odours, since agreeableness and the right degree belongs to the mean. In the vegetables pruning is not possible, except with those that improve when cut off at the ground, as the leek and any that will sprout repeatedly.[3] But the removal of other plants growing with them, and indeed of plants of the same kind when crowded together, is in effect a sort of pruning, since it provides the remainder with more food and lets them grow more.

19.3

With some of these plants special measures are also taken, as we said earlier[4] of rue, and as is true of others as well.

But there are fewer such special cases here, and the reasons (one may say) are evident.

Husbandry and Seed-Crops

What remains are matters that are in contrast

20.1

---

cf. also *CP* 2 15. 6.

[3] Cf. *HP* 7 4. 10 (of kinds of onion): "The kind called *géteion* (horn-onion) . . . is often cut down, like the leek . . ."; *cf.* also *CP* 2 15. 6.

[4] *CP* 3 17. 1 (ashes used with fig and rue); compare also perhaps *CP* 2 5. 3 (watering with saline water improves beet, rue and rocket among vegetable-like plants).

THEOPHRASTUS

τὰ[1] δένδρα, τὰ περὶ τὴν ψιλὴν γεωργίαν ἐστίν· ἐνταῦθα δὲ ἥ τε τῆς γῆς κατεργασία, καὶ ἡ τῶν σπόρων ὥρα, καὶ μετὰ ταῦτα σπαρέντων κατεργασία, πρότερον δ' ἔτι[2] τούτων, ἢ καὶ ἅμα, τὴν οἰκείαν ἑκάστῳ <καὶ>[3] σπέρμασιν[4] ἰδεῖν, ὥσπερ καὶ ἐπὶ τῶν δένδρων.

ἔστι δὲ καὶ κατὰ τὰς ἐργασίας τὸ οἰκεῖον, οἱονεὶ τὴν μὲν θέρους μᾶλλον, τὴν[5] δὲ χειμῶνος, νεᾶν ἢ σκάπτειν ἤ τι τοιοῦτον ἕτερον, ἅπερ ἐπιχειροῦσί τινες διαιρεῖν. δεῖ γὰρ ὥς φασιν τὴν μὲν ἔπομβρον καὶ στερεὰν καὶ βαρεῖαν, καὶ τὴν πίειραν, θέρους ἐργάζεσθαι καὶ τοῖς ἀρότοις καὶ τῇ σκαπάνῃ, τὴν δὲ ξηρὰν καὶ μανήν, καὶ τὴν λεπτὴν καὶ κού-

20.2

---

[1] u : τοU.
[2] u aP : δέτη U : δέ τι N.
[3] ego.
[4] U : σπέρματι u.
[5] τὴν ... τὴν Wimmer : τῇ ... τοῦ U N : τοῦ ... τοῦ aP.

---

[1] Of the great classes of plants the husbandry of trees (*CP* 3 2. 6–3 18. 5), of undershrubs (with coronaries, which are partly undershrubs and partly herbaceous plants) and vegetables (*CP* 3 19. 1–3) has been treated. The remaining class is that of herbaceous plants (*CP* 3 20. 1–3 24. 4),

DE CAUSIS PLANTARUM III

(as it were) to the cultivation of trees,[1] to wit the matters concerned with the husbandry of field crops.[2] Here belong the tillage of the ground, the season for the sowing and after these the care of the plants when sown—but coming before these (or else contemporary with them)[3] is seeing with grains too (as we saw[4] with trees) what country is appropriate for each sort.

The question of the appropriate country also applies to differences in tillage,[5] for instance whether to plough or spade or treat in some other such way one kind of country rather in summer, another in winter, distinctions that certain agriculturists endeavour to draw. So they say that country 20.2 that is rainy and compact and heavy, and fat soil, should be worked in summer, both by ploughing and spading, whereas dry and open-textured country,

---

which includes the "seeds" or grains, that is, legumes and cereals.

[2] "Field crops" renders *psilè geōrgía*, literally "bare farming," that is, the farming of land that is bare (*psilè gê*; *cf. CP* 3 6. 5) of trees. the phrase therefore contrasts tree-farming with field-farming. Gardening and "summer seeds" are not considered.

[3] Certainly not *after* them.

[4] *CP* 2 4. 4–5, 7–8.

[5] And not simply to the land sown.

THEOPHRASTUS

φην, καὶ¹ τοῦ χειμῶνος · δύναται γὰρ ἡ μὲν ξηραίνειν καὶ λεπτύνειν, ἡ δὲ χειμερινὴ παχύνειν καὶ ὑγραίνειν, ἑκατέρα² δὲ τούτων δεῖται πρὸς τὸ ἐνδεὲς τῆς φύσεως.

καὶ κοπρίζειν πλεῖον μὲν τὴν λυπράν, ἔλαττον δὲ τὴν ἀγαθήν, καὶ διὰ τὴν ἀρετὴν τῆς γῆς, καὶ διότι τῷ³ διὰ τὴν κόπρον⁴ πλείω λαμβάνειν τροφὴν λοχαῖος πίπτει.⁵

20.3 μίσγειν δὲ καὶ τὴν γῆν τὴν ἐναντίαν, οἷον τῇ βαρείᾳ τὴν κούφην, καὶ τῇ κούφῃ τὴν βαρεῖαν, καὶ τὴν λεπτὴν τῇ πιείρᾳ · ὡσαύτως δὲ καὶ τὴν ἐρυθρὰν καὶ τὴν λευκήν, καὶ εἴ τις ἄλλη ἐναντιότης. οὐ γὰρ μόνον ἡ μῖξις ἀποδίδωσι τὸ ἐλλεῖπον, ἀλλὰ καὶ ὅλως σφοδροτέρας ποιεῖ, καὶ ἐὰν <ἡ>⁶ τυχοῦσα ᾖ⁷ μῖξις · οἷον ἐὰν τὴν ἀπειρηκυῖαν καὶ μὴ δυναμένην φέρειν ἑτέρᾳ μίξῃς,⁸ πάλιν φέφει καθαπερανεὶ καινὴ γεγενημένη, καὶ ἡ καθ' αὑτὴν οὐ φέρουσα, καθάπερ ἡ ἄργιλος, ὅταν μιχθῇ, ποιεῖ φορόν ·⁹ ὥσπερ γὰρ κόπρος ἡ ἑτέρα τῇ ἑτέρᾳ
20.4 γίνεται. διὸ ταύτῃ¹⁰ ποιοῦσι¹¹ Μεγαρεῖς,¹² καὶ

---

¹ [καὶ] Gaza, Scaliger.   ² u : ἑκάτερα (ἐ- U) N aP.
³ ego : διατὸ U.
⁴ τὴν κόπρον U : ἐκ τῆς κόπρου Schneider.
⁵ λοχαῖος πίπτει (sc. ὁ σῖτος) ego (ἢ ὥστε πέττειν Itali after Gaza : ἢ λοχαῖος πέττειν Scaliger) : λεχέος πίττῃ U.

DE CAUSIS PLANTARUM III

and soil that is lean and light, should be worked even in winter, since summer working has the effect of making the soil drier and leaner, whereas winter working makes it thicker and wetter, and each soil needs these qualities to offset what is lacking in its nature.

Again one should (they say) manure poor country more but good country less, both because the soil is excellent in good country and because the grain, by taking too much food on account of the manure, gets heavy-headed and lodges.

They also tell us to mix the soil with its opposite: thus we should mix light soil with heavy, heavy with light, fat with lean, and so too with white and red soil and so with other opposites. For the mixture does not merely give the soils what they lack but in fact makes them more vigorous in general, even if the mixture is not with the opposite. So if one takes soil that is worn out and unable to bear and mixes it with another, it bears once more, as if it had become new again; and soil that taken alone does not bear, as clay, when mixed with such worn-out soil, makes it productive, since the one turns out to be (as it were) manure to the other. Hence the

20.3

20.4

---

[6] Schneider.   [7] ἢ aP : ἡ U N.
[8] μίξης u^c (-εις u^{ac}) a : μυξῆς U : μίξις N P.
[9] Schneider : φόρον U.   [10] ταυτη U : ταῦτα u.
[11] U^c : ποιοῦσοι U^{ac} (ut vid.).   [12] u : μεγάρης U.

THEOPHRASTUS

δι' ἐτῶν πέντε ἢ ἓξ σκάπτοντες βαθεῖαν, ἀναβάλλουσιν τὴν κάτωθεν ἐφ' ὅσον δικνεῖται τὸ ὕδωρ ἵνα καινὴν ποιῶσιν· ἡ γὰρ τρόφιμος ἀεὶ καταρρεῖ ὑπὸ τοῦ ὕδατος.

σπείρειν δὲ κελεύουσι τὴν ξηρὰν καὶ τὴν θερμὴν <πρὸ>[1] τῶν ὑδάτων, ὅπως λαμβάνουσα πλῆθος ὕδατος ἐκτρέφειν δύνηται· καὶ τὴν ἐπομβρίαν δ' ὡσαύτως, ἵνα τὰ σπέρματα λαμβάνωσι,[2] καὶ βλαστόντα[3] καὶ αὐξηθέντα θερμῇ τῇ γῇ δυνατώτερα δέχηται τὸν ὄμβρον· ἐὰν δὲ μή, ἐν ξηρᾷ, προεργασάμενον[4] τὴν γῆν, ὀψὲ ποιεῖσθαι τὸν σπόρον. 20.5 σπείρειν δὲ δεῖ τὸν μὲν πρώϊνον μανόν, τὸν δὲ ὄψιον πυκνόν·[5] ὁ μὲν γὰρ οὐκ ἔχει ῥίζωσιν, ὁ δ' ἔχει, καὶ γίνεται πολυσχιδής.[6]

ταῦτα μὲν οὖν καὶ τὰ τοιαῦτα (πλείω γάρ ἐστι), διακριβοῦν εἴ τις ἐθέλει τὰς ἐργασίας μᾶλλον· ὧν[7] ἕκαστα δεῖ σκοπεῖν πρὸς τὴν χώραν. ἔνιαι[8] γὰρ οὐ δέχονται τὰς ἀκριβείας, ἀλλὰ δι-

---

[1] Gaza (*ante*), Itali.
[2] N aP : λαμβάνουσι U : λαμβάνουσα u : Schneider deletes (but for the sense "take hold" *cf. HP* 6 2. 4, 6 2. 6).
[3] U : βλαστοῦντα u.
[4] Gaza, Schneider : πρὸς ἐργασάμενος U.
[5] u aP (ὄψιο πυκνὸν N) : ὀψιόπυκνον U.
[6] u : -εδής U.   [7] [ὧν] Gaza, Schneider.
[8] Gaza, Wimmer : ἔνια U.

DE CAUSIS PLANTARUM III

Megarians act on these lines, and every four or five years spade deep and turn up the bottom soil as far down as the rain reaches, in order to make the soil new, since the nutritious soil is constantly carried down by the rainwater.

As for sowing, we are told to sow dry country and hot country before the rains,[1] so that the country, getting plenty of rainwater, may be able to bring up the plants; and similarly with rainy country, so that the seeds may take, and once they have sprouted and grown in the earth while it is warm, may receive the rains with greater staying power. Failing that, one is to make one's tardy sowing[2] when the ground is dry, first working the soil over. In the early sowing we must sow the grain thin, in the late sowing thick,[3] since the late grain does not root well, whereas the early does, and comes up with many branchings.

These then and the like (for there are more) are the precepts if one wishes to enter into the refinements of tillage. Each must be examined with reference to the country, since some countries reject

---

[1] The rains that follow the setting of the Pleiades (Nov. 6–9); *cf. CP* 3 23. 1.
[2] *Cf.* Hesiod, *Works and Days*, 479–90 on the consequences of missing the proper season for sowing.
[3] The early sowing is in autumn, the late in spring. The roots of grain sown in spring do not have the benefit of winter (*cf. CP* 2 1. 4).

THEOPHRASTUS

ἁμάρτοι ἂν ὁ ποιῶν, ὅπερ ἔπαθέν τις ἀνελθὼν εἰς Συρακούσας ἐκ Κορίνθου τὴν ἐργασίαν τὴν ἐντεῦθεν μεταφέρων. καὶ ὃς[1] τὸ χωρίον τὸ δοκοῦν ἀγαθὸν ἐκλιθολογήσας διέφθειρεν (ἐξεπήγνυτο[2] γὰρ ὁ σῖτος οὐκ ἔχων οὐδεμίαν προβολὴν ἐξαιρεθέντων τῶν λίθων) · καὶ ὃς[3] ἐν Συρίᾳ, κατὰ βάθος[4] ἀρόσας, ἀνεξήρανεν[5] (ὑπὸ γὰρ τοῦ θέρους διακαίοντος[6] ἐπὶ πλεῖον ὑπόπετρος οὖσα διέκαιε · <διὸ>[7] καὶ μικροῖς ἀρότροις[8] οἱ Σύροι χρῶνται).

[διὸ][9] ταῦτα μὲν πειρατέον ἐπικρίνειν.

20.6 ὡς δὲ κοινῇ[10] καὶ καθόλου πᾶσιν[11] εἰπεῖν.

---

§ 5 lines 8–16 Pliny, N. H. 17. 30: at in Syracusano agro advena cultor elapidato solo perdidit fruges luto, donec regessit lapides. In Syria levem tenui sulco imprimunt vomerem, quia subest saxum exurens aestate semina.

[1] ego (οὕτως Dalecampius : γὰρ Wimmer) : ὡς U.
[2] U^r (-υντο U^ar) : ἐξεπίγνωτο N^c (from -ἐ-) aP.
[3] ego (καὶ ὁ Schneider : καί τις Wimmer) : καὶ ὡς U.
[4] U : βάθους W. Müller.
[5] u : ἀρόσασαν ἐξήρανεν U N aP.
[6] ego (διακαιομένη <ἡ γῆ> Schneider) : διακαιομένης U.
[7] Schneider (unde fit ut Gaza).
[8] aP : ἀρότοις U N.
[9] Schneider.
[10] κοινῇ u aP : -ὴ U N.

the refinements and one who applied them would fail. This was the experience of a man who returned[1] to Syracuse from Corinth and introduced the practices of those parts. This man ruined a farm of good repute by clearing out the stones, for the grain froze when the removal of the stones left it without protection from the cold. Another man in Syria dried his crop out by ploughing deep,[2] for the soil was now more deeply scorched by the summer heat, and having stones in it scorched the crop (which is why the Syrians use small ploughs).

These are refinements which we must endeavour to pass under review.

Seed Crops: Ploughing

But speaking on the general level and including all sown crops, the thing of greatest importance and   20.6

---

[1] Perhaps in consequence of Timoleon's invitation to exiles and others to settle in Syracuse (Plutarch, *Timoleon*, chap. xxiii 1–3; about 343 B.C.).
[2] *Cf. HP* 8 6. 3 (of sown crops): "In some countries they say that it is harmful to plough deep, as in Syria, which is why they use small ploughs. Elsewhere cultivation when pushed to extremes does harm, as in Sicily, which is why many foreigners, we gather, fail there."

---

[11] Wimmer (Gaza omits : Schneider deletes) : φασὶν U.

THEOPHRASTUS

μέγιστον μὲν καὶ πρῶτόν ἐστι τὴν[1] σπορευτὴν[2] χώραν κατειργάσθαι[3] καλῶς, εἰς διειργασμένην γὰρ καλῶς πεσὸν τὸ σπέρμα, καὶ ἐκφύεται, διημερωθείσης τῆς γῆς· εἶθ' ἡ κόπρισις ἤδη καὶ ἄλλη καὶ ἄλλη θεραπεία μετὰ ταῦτα διαβεβλαστηκότων, οἷον σκάλσις καὶ ποασμός.

20.7 ἡ κατεργασία δὲ[4] ἐν τῷ νεᾶν κατ' ἀμφοτέρας τὰς ὥρας, καὶ θέρους καὶ χειμῶνος, ὅπως χειμασθῇ καὶ ἡλιωθῇ ἡ γῆ (καθάπερ καὶ ἐπὶ τῆς φυτείας ἐλέχθη)· πολλάκις γὰρ μεταβληθεῖσα, μανὴ καὶ κούφη φαὶ καθαρὰ γίνεται τῆς ὕλης, ὥστε ῥᾳδίως ἐκτρέφειν. καὶ διὰ τοῦτο[5] κελεύουσιν οὐδὲ τὰ χεδροπὰ συμβάλλειν εἰς τὰς νεούς[6] (ἐὰν μή τι[7] σφόδρα πρώϊον), ὅπως μὴ κωλύωσιν τὴν θερινὴν νέανσιν. ἀγαθὴν δὲ οἴονται τὴν χιόνα ταῖς χειμεριναῖς νεοῖς, καὶ οὐχ ἧττον τὴν πάχνην εἶναι,

20.8 διεσθίειν[8] γὰρ καὶ μανοῦν τὴν γῆν. καὶ ὅταν μετὰ

---

[1] Wimmer (*primum est* Gaza: πρῶτον τὸ τὴν Schneider): πρῶτον εἴ τις U : πρῶτον εἴ τι N : πρῶτον εἰ τὴν aP.
[2] R, Stephanus : πορευτὴν U.
[3] u P : κατειργάσθαι U : κατείργασθαι N : πατείργασται a.
[4] ego : δὲ κατεργασία U.
[5] U : τοῦτ' οὐ ? ego.
[6] ego : νέας U.
[7] Schneider : τις U.
[8] u aP : διαισθίειν U : διευσθίειν N.

DE CAUSIS PLANTARUM III

coming first is that the land to be sown should have been well worked,[1] for when the seed falls on thoroughly well-worked land it also comes up, since the soil has been thoroughly tamed. Next comes manuring and the various operations after these, when the seeds have sprouted, such as hoeing and weeding.

20.7 The working consists in ploughing in both seasons, both in summer and in winter, so that the soil may be exposed to winter and to the sun, a point we also made[2] in treating the planting of trees. For by being turned up often the soil becomes open-textured, light and free of woody plants, so that it can easily bring up the crop. For the reason why we are told not even to put legumes[3] in the ploughed ground, unless the legume is very early,[4] is that they may not interfere with the summer ploughing. Snow is considered excellent for fields ploughed in winter, and hoar-frost no less, for (they say) it eats through the ground and gives it an open texture.[5] Again when farmers after the first 20.8

---

[1] *Cf. HP* 8 7. 7 (on the importance of various factors for seed crops): "The working of the land contributes also in no small measure and especially the working for the sowing, since when properly worked over the land easily brings forth the crop."   [2] *CP* 3 4. 1.

[3] Legumes renew the soil: *cf. HP* 8 7. 2, 8 9. 1; *CP* 4 8. 1.

[4] That is, produces a crop soon after sowing.

[5] *Cf. CP* 5 12. 7. For snow giving an open texture *cf. CP* 3 23. 4.

τοὺς πρώτους ἀρότους νεάσωσιν πάλιν τοῦ ἦρος μεταβάλλουσιν ὅπως τὴν ἀναφυομένην πόαν ἀπολέσωσιν, εἶτα θέρει ἀροῦσι, καὶ πάλιν ὅταν μέλλωσι σπείρειν ὑπήροσαν, ὡς δέον (καθάπερ ἐλέχθη) προκατεργάσασθαι, καὶ περὶ τούτου μάλιστα σπουδάσαι· διὸ καὶ τὴν ἀπὸ τῆς δικέλλης ἐργασίαν μᾶλλον ἐπαινοῦσιν, ἡ δ' ἀπὸ τοῦ ἀρότρου δοκεῖ πολλὰ παραλείπειν.[1] Θετταλοὶ δ' ἰσχυρότερον ἔτι τῆς δικέλλης ὄργανον ἔχουσιν, ὃ καλοῦσιν μίσχον, ὃ μᾶλλον εἰς βάθος κατιὸν πλείω γῆν περιτρέπει καὶ κατωτέρωθεν.

20.9 αὕτη μὲν οὖν τῆς χώρας ἡ κατεργασία· συμβαίνει δὲ τοῖς εἰς ταῦτα πονήσασιν ἧττον ἐν τοῖς ἄλλοις κακοπαθεῖν· ὅσοι δ' ἂν μαλακωτέρως, ἀναγκάζονται πολλάκις καὶ σκάλλειν καὶ βοτανίζειν, ἅτε πολλῆς πόας ἐκφυομένης. οὕτω[2] γὰρ ἀπόλλυται τὰ σπέρματα, προτερεῖ δὲ τοῦ σίτου διά τε τὴν ἰσχὺν καὶ τὸ προϋπάρχειν· ἐπεὶ πρότερον αὐξηθέντος τοῦ σίτου πολλὰ καταπνίγεται καὶ

---

[1] u : παραλειπεῖν U : παραπεῖν N : παραποιεῖν aP.
[2] Gaza, Basle ed. of 1541 : οὔτε U.

---

[1] "First ploughing" is also used of winter ploughing at *HP* 8 6. 1; at *HP* 8 1. 2 "ploughing" without qualification is

ploughing[1] plough again in spring they turn the earth to destroy the weeds that come up, and then plough in summer and plough lightly once more just before sowing,[2] with the idea (as we said)[3] that one must work the land before sowing and make this one's chief task. This is why the authorities prefer working the land with the mattock, and consider that working it with the plough omits a good deal. The Thessalians have a tool still more effective than the mattock, which they call "mischos," which by going deeper turns up more soil and from further down.

This then is the working of the country. Men who have worked hard at these tasks find that they spare themselves labour in the rest, whereas those who have exerted themselves less are forced to hoe and weed again and again, since there is a plentiful growth of weeds. For this is how their seeds are destroyed before they can develop, and the seeds come out before the grain both on account of their vigour and because they are in the ground first. Whereas if the grain comes up first many of the weeds are

20.9

---

used of it.

[2] That is, the sowing at the setting of the Pleiades (Nov. 6–9); cf. *HP* 8 1. 2: "The seasons of sowing are for most of them (*sc.* field crops) two: the first and main season is that at the setting of the Pleiades ...; there is another at the beginning of spring after the winter solstice."

[3] *CP* 3 20. 6.

ἀπόλλυται, καὶ τὸ ὅλον οὐχ ὁμοίως βλάπτει τὰ σπέρματα.

21.1 τὰ δ' οἰκεῖα τῇ μὲν ἐπόμβρῳ καὶ ψυχρᾷ τὰ ἀνοστιμώτατα[1] πρὸς τὴν σιτοποιίαν, ὁμοίως καὶ πυρῶν καὶ κριθῶν καὶ τῶν ἄλλων· ἀραιὰ γὰρ ταῦτα τὰ δ' ἅμα[2] πλείονος ὄμβρου δεῖται· καὶ τὰ μέλανα τῶν λευκῶν. ἐὰν δ' ἐν[3] ξηρᾷ καὶ ἐρυθρᾷ καὶ ἐν ψυχροῖς τόποις, παλίνοστα·[4] καὶ τὰ λευκὰ μᾶλλον τῶν μελάνων, ὅπως ἁδρύνηται πρὸ τῆς ἀπολείψεως τῶν ὑδάτων· αὔξεται γὰρ καὶ φθάνει[5] τὰ λευκὰ μᾶλλον τῶν μελάνων. ἐν δὲ ταῖς εὐκρασίαις τὰ ἀνὰ μέσον. οἱ μὲν οὖν οὕτω διαιροῦσιν, οἱ δ' ἄλλως.

21.2 ὡς δ' ἁπλῶς εἰπεῖν ἡ μὲν λεπτὴ[6] κριθοφόρος ἀμείνων, ἡ δὲ πίειρα πυροφόρος· αἱ μὲν γὰρ ἐλάττονος καὶ κουφοτέρας δέονται τροφῆς, οἱ[7] δὲ

---

[1] Schneider : ἀναστομωτατα U^c from ἀναστωματάτα.
[2] τὰ δ' ἅμα ego (καὶ Schneider) : τα δ' αλλα U.
[3] ἐὰν δ' ἐν ego (ἐν δὲ Schneider) : ἐὰν δε U.
[4] U : τὰ πολύνοστα Schneider : πολύνοστα Wimmer.
[5] Gaza (*celerius*), Schneider : φθίνει U.
[6] u : λεπτᾶν U.   [7] Schneider : αἱ U.

---

[1] The volume of barley meal or wheat flour was compared to the volume of the barley corn or wheat kernel

DE CAUSIS PLANTARUM III

choked and destroyed, and in general their seeds do less harm.

Seed Crops: Varieties and Country

The cereals that are suited to rainy and cold country are the varieties with kernels having the lowest yield in meal,[1] alike in wheat, barley and the rest, since these kernels are open in texture and with this character goes a need for more rain; again the black-grained varieties require more than the white.[2] Whereas if one sows on dry and red ground and in cold districts the crops to sow have an equal yield in meal; and the white-grained variety is to be sown in preference to the black, so that the plant may become sturdy before the end of the rains, since the white grows and gets sturdy sooner. Where the land has an equable blend of qualities the intermediate varieties are to be sown. Some authorities, then, distinguish as above, others differently. 21.1

On the whole lean ground is a better producer of barley, rich ground of wheat, barley requiring less 21.2

before grinding.

[2] *Cf. HP* 8 4. 2 (of barleycorns): "Further, some are white, others black with a reddish tinge, and these are considered both to yield a large amount of meal and to be stronger in resisting cold and wind, in short the weather, than the white." For wheat Theophrastus gives no such particulars about colour, but merely says that there are differences (*HP* 8 4. 3).

THEOPHRASTUS

πλείονος καὶ σωματωδεστέρας.

τῶν δὲ πυρῶν ὁ μὲν τρίμηνος ἐν ταῖς[1] λεπτογείοις καλλίων, σύμμετρος γὰρ ἡ τροφὴ κούφη κούφοις·[2] ὁ δὲ Λιβυκὸς και ὁ Δρακοντίας καὶ ὁ στλεγγίας[3] καὶ ὁ Σελινούσιος ἐν ἀγαθῇ, πολύτροφοι γάρ (σημεῖον δ᾽, ὅτι κάλαμον ἔχουσι παχύν·[4] ἁπλῶς δὲ καὶ εἴ[5] τις ἄλλος τοιοῦτος). εἰ δέ τις πολυχίτων, ὥσπερ ὁ Θρᾴκιος, ἐν ταῖς χειμεριναῖς, ἀπαθὴς γὰρ ὑπὸ τοῦ ψύχους· ὁ δὲ χαῦνος καὶ μανός, ὥσπερ ὁ καχρυδίας,[6] ἐν ταῖς[7] ἐπόμβροις, τροφῆς γὰρ πολλῆς καὶ οὗτος δεῖται (διὸ καὶ παχυκάλαμος).

21.3 ὡσαύτως δὲ καὶ ἐπὶ τῶν χεδροπῶν. ἡ μὲν γὰρ γλίσχρα καὶ μελάγγεως ἐρέβινθον, ἡ δὲ κούφη

---

§ 2 lines 7–9 Pliny, *N. H.* 18. 64: ex omni autem genere grani praetulit (*sc.* Graecia) dracontian et stlengian (stelepan *or* istelepan MSS.) et Selinusium argumento crassissimi calami.

lines 13–15 Pliny, *N. H.* 18. 69: plurimis tunicis Thracium triticum vestitur ob nimia frigora illi plagae exquisitum.

[1] Schneider : τοῖς U.   [2] u : -ως (-ους?) U.
[3] ego (stlencias *or* stlentias Gaza) : στρεγγιας U (at *HP* 8 4. 3 U twice has στλεγγύς).   [4] U^a : παχύ U.
[5] u : ἤ U.   [6] ego (so U twice at *HP* 8 4. 3) : καχυδριας U.
[7] Schneider : τοῖς U.

DE CAUSIS PLANTARUM III

and lighter food, wheat more and heavier.

Of the varieties of wheat the three-months kind is finer in country with lean soil, for light food is just right for light wheat. But the Libyan,[1] Dracontias,[2] stlengias[3] and Selinuntine varieties are finest in good soil, since they need plenty of food (proof of this is their thick haulm, and this is true of all thick-haulmed varieties). Many-coated wheat, on the other hand, like the Thracian,[4] grows finest in wintry countries, since it is not affected by the cold; spongy and open-textured, like the cachrydias,[5] in rainy country, since it too needs plenty of food (which is why it has thick haulms).

So too with legumes: viscous and black soil produces finer chick-peas,[6] but light soil finer beans, 21.3

---

[1] *Cf. HP* 8 4. 3: "And some wheats have a thin, some a thick haulm; this too the Libyan wheat has, and the cachrydias has a thick one too."

[2] Etymology unknown; perhaps from *drákōn* (serpent), because it was spotted or resembled one of the plants called *drakóntion* ("snake-weed"), or perhaps from the name Dracon.

[3] From *stlengís* ("strigil"), no doubt because it was thick and curved at the end like the handle of a cane.

[4] *Cf. HP* 8 4. 3: "And further some wheats have few coats, some many, like the Thracian."

[5] From *káchrys* ("parched barley"), which the grain may have resembled.

[6] *Cf. HP* 8 7. 2 (of chick-pea): "In a word no ordinary soil can bear it, but to do so the soil must be black and fat."

THEOPHRASTUS

κύαμον φέρει καλλίω, σύμμετρα γὰρ ἑκατέρῳ τὸ τῆς <γῆς>.¹

καὶ ἐπὶ τῶν ἄλλων δὲ τὸν αὐτὸν τρόπον, ἀεὶ πρὸς τὰ τῆς τροφῆς ἀνακτέον, συμπαραλαμβάνοντα καὶ τὰ² τοῦ ἀέρος. ἡ γὰρ εὔειλος (καὶ ἁπλῶς ἡ εὐδιεινή) τὰ ἀσθενέστερα συμφέρει³ μᾶλλον, ἡ δ' ἐναντία τὰ ἰσχυρά. καὶ ἡ ὕφαμμος⁴ δὲ καὶ μηλώδης⁵ τὰ ὀλιγότροφα καὶ ξηρά, καθάπερ τὴν Ἀχιλληΐδα,⁶ καὶ γὰρ λευκαίνει, μᾶλλον ἐκπέττουσα τὰς τροφάς · καὶ φέρει δὲ καὶ τεράμονα καὶ⁷ ξηρὰ καὶ τὰ πρόσειλα, πέττει γὰρ ὁ ἥλιος μᾶλλον, ἐν δὲ ταῖς⁸ εὐγείοις καὶ πίοσιν οὐχ ὁμοίως.

21.4 πυρὸς δὲ κριθῆς ἐν τοῖς⁹ ὀμβρώδεσιν εὐθενεῖ μᾶλλον, καὶ τὸ ὅλον δὲ πρὸς τὰς ἐπομβρίας ἰσχυρότατος. φέρει δὲ καὶ ἐν τοῖς ἀκόποις μᾶλλον · αἴτιον δὲ ὅτι θερμὸς ὢν ἐκπέττειν δύναται μᾶλλον

---

¹ ego (*utrique ... alimentum offertur* Gaza : ἑκατέρῳ τὰ τῆς <τροφῆς> Heinsius) : ἑκατερωτάτης U.

² U^c : τὰς U^{ac} N aP.

³ U : φέρει Scaliger : ἐκφέρει Wimmer.

⁴ Scaliger : ἀνύφαμμος U^c (φα U^{cm} with an index over the right leg of the first μ) : ἀνύμμος U^t.

⁵ ego (Schneider deletes or reads καὶ <οὐ> πηλώδης : καὶ ἁλμώδης Hindenlang) : καὶ πηλώδης U.

⁶ U^c (ΐ ss.).   ⁷ U N : τὰ aP.

DE CAUSIS PLANTARUM III

since the provision in the soil is right for each.

So too with the rest: we must always account for the country by the food, also including with the food the character of the air. For country that is sunny (in short that has fair weather) tends more to bring with it the weaker plants, whereas the opposite sort of country favours the strong. Again country with sand below and yellow does better with plants that take little food and are dry, like the Achilles barley,[1] which indeed turns white, concocting its food more thoroughly. Again places too that are in the sun bear grains that are tractable and dry, since the sunlight concocts them better, whereas this happens less in rich and fat land.

Wheat does better than barley in rainy regions 21.4 and is in short most resistant to rains. It also bears better than barley on land that is not manured. The reason is that wheat is hot and so better able than barley to concoct its food thoroughly and does not

---

[1] For this barley *cf. HP* 8 4. 2; *HP* 8 10. 2; *CP* 3 22. 2 (where both a white and a black variety are mentioned). *Cf.* Galen, *Linguarum seu Dictionum Exoletarum Hippocratis Explicatio* (vol. xix, p. 87 Kühn): "Achilles barley <Hippocrates, *Diseases III*, 17 (Loeb vol. V, p. 58)>: plump and large barley, named it is said from one Achilles, a farmer of Brauron (? Βαβρωνίου)."

---

[8] Schneider : τοῖς U.

[9] U : ταῖς Schneider.

THEOPHRASTUS

καὶ κόπρον οὐχ ὁμοίως ἐπιζητεῖ διὰ τὸ ἔχειν ἐν
ἑαυτῷ ὃ ἔδει ἐκ τῆς κόπρου γίνεσθαι, καὶ τὸ ὅλον
ἰσχυρότερός ἐστι καὶ πρωϊσπορεῖται καὶ ἐν τοῖς
ψυχροῖς καταβάλλεται μᾶλλον. ἡ δὲ ἰσχὺς θερμότητι καὶ τῷ μᾶλλον πολυχίτωνα εἶναι· διὸ καὶ ἡ
διὰ χρόνου γεωργουμένη γῆ πυροὺς φέρει μᾶλλον
ἢ[1] κριθάς· κατακρατεῖ γὰρ τῆς τροφῆς ἰσχυροτέ-
ρας οὔσης ὁ πυρὸς μᾶλλον, ἐπεὶ οὐχ[2] ὁμοίως ἐθέλει[3] φέρειν μὴ ἐμπρησθείσης[4] τῆς ὕλης· οὕτως

21.5 δ' ὥσπερ μανοῦται καὶ κοπρίζεται.[5] ὡς δὲ ἁπλῶς
εἰπεῖν ἐν τρισὶ[6] ταῦτα[7] μεγίσταις καὶ κοινοτά-
ταις ἐστὶ διαφοραῖς· ἐν τῇ τῆς χώρας φύσει
<καὶ>[8] τῇ τῶν σπερμάτων ἰσχύϊ καὶ ἀσθενείᾳ
καὶ ἐν τῇ τοῦ ἀέρος κράσει, διὸ σκεπτέον ταῦτα.

χειμῶνες δὲ ἐπιγινόμενοι πανταχοῦ μὲν χρήσι-

---

[1] u : καὶ U N aP.
[2] ego (ἐνιαχοῦ δ' οὐχ Wimmer) : ἐπέχει δ' οὐ U.
[3] <δ'> ἐθέλει Schneider.
[4] Schneider : ἐμπλησθείσης U.
[5] δ' ... κοπρίζεται U : γὰρ μ. καὶ ὥσπερ κ. <ἡ γῆ> Schneider.   [6] U : πᾶσι N : πάσαις aP.
[7] ego : ταύταις U.   [8] ego.

---

[1] That is, heat.
[2] That is, in autumn instead of spring. Barley is in fact sown earlier than wheat (*HP* 8 1. 3).

DE CAUSIS PLANTARUM III

seek manure to the same extent because it possesses within itself the thing[1] that the manure was to provide; in a word it is stronger and is sown in the early season[2] and in cold regions more than barley.[3] The strength lies in its heat and in its having coats[4]; and this is why ground worked after lying fallow for some time is a better producer of wheat than of barley,[5] since wheat masters the food, which here is stronger, better. (Indeed such ground refuses to bear so well if the woody plants have not been burnt over; but when this is done the soil is as it were loosened and manured.)[6] All this[7] in a word comes under three greatest and most universal variables: the nature of the land, the strength or weakness of the plants, and the tempering of the air, which is why these are the points to study.

21.5

Ensuing cold weather is everywhere beneficial,

---

[3] *Cf. HP* 8 6. 4: "... in wintry country they (*sc.* the agriculturists) tell us to sow wheat rather than barley, and in general cereals rather than legumes in land cultivated after a long interval, these too bearing wheat better than barley. Wheat can also resist rainy weather better than barley, and bears better in unmanured land."

[4] *Cf. HP* 8 4. 1: "Then too wheat is in a number of coats but barley is naked, for barley has the most naked seed of the cereals."

[5] *Cf.* note 3 above.

[6] The burning, like manuring (*cf. CP* 3 6. 1), produces openness of texture and heat.

[7] The question of what grains do better in what country.

163

THEOPHRASTUS

μοι, ῥιζοῦται γὰρ καὶ καρκινοῦται μᾶλλον, ὥστ᾽ εἰς τὸ ἔαρ σύμμετρον ἀποδίδοται μέγεθος · [1] μάλιστα δ᾽ ἐν ταῖς ἀγαθαῖς χώραις, τὰ γὰρ νότια καὶ ὅταν εὐδία ταχὺ ἀναδιδόασιν καὶ ποιοῦσι λοχαῖον [2] καὶ τοῦ ἦρος [3] ἐρυσιβώδη.

τὰ δὲ πυκνόσπορα πρότερον ἀποχεῖται τῶν μανοσπόρων ὅτι τὰ μὲν ῥιζοῦται καὶ εἰς τὰ κάτω σχίζεται διὰ τὸ ἔχειν τόπον, τὸ δ᾽[4] εὐθὺς εἰς τὸ ἄνω τρέχει.

22.1 χρὴ δὴ[5] καὶ τὰ οἰκεῖα σπέρματα πρὸς εὐφορίαν καὶ τὰ νοσήματα τὰ συμβαίνοντα λαμβάνειν, οἷον ἐρυσίβας, ἐπεὶ <καὶ>[6] ἐν τοῖς κοίλοις καὶ ἀπνόοις γίνεται[7] τὰ μὴ ἐρυσιβούμενα · τοιαῦτα δὲ τὰ ἐπικλινῆ τῷ στάχυϊ καὶ μὴ ὀρθά. διὰ τοῦτο δὲ καὶ τὸ ἐπικύπτειν συμφέρει τὸν στάχυν, ὅπως ἀπορρέῃ καὶ μὴ ἐμμένῃ τὸ ὕδωρ καὶ ἡ δρόσος · [8] ἀποκλίνουσι δὲ οἱ μακροὶ στάχυες μᾶλλον, οἱ δὲ πλατεῖς καὶ βραχεῖς ὀρθότεροι, διὸ καὶ ἐρυσιβοῦνται. ταῦτα δὲ

---

[1] <τὸ> μέγεθος Schneider (cf. CP 3 23. 5).
[2] ego : λοχαῖαν U.
[3] ego ([τοῦ] πρὸς Scaliger : καρπὸν Wimmer) : τοῦ πρὸς U.
[4] U (τόδ᾽ N aP) : τά δ᾽ u.   [5] U : δὲ N aP.   [6] ego.
[7] γίνεται <τὰ ἐρυσιβούμενα, ἐν δὲ τοῖς εὐπνόοις> Wimmer[c].
[8] ἡ δρόσος u : ἰδρόσος U.

for the grain then gets better rooted and tillered,[1] so that in spring it grows to no more than the right height. The cold is most useful in good country, for southerly winds, even with clear skies, make the grain shoot up too rapidly and become lodged and get rust in spring.

Thick-sown grain heads earlier than thin-sown, since thin-sown goes to root and branches out downwards below ground, because it has the room, whereas the thick-sown runs upward at once.

Seed Crops: Diseases (1) Rust

So it is also the appropriate *grain* that we must choose to obtain a good crop and to avoid the diseases that occur, such as rust, since even in hollow and windless places we have types that do not rust.[2] Of this sort is grain with ears that incline and do not stand erect. The reason why it is also[3] good that the ear should bend is that the ear should let the rain and dew flow off and not remain on it. Long ears do more bending, but broad and short ears stand straighter and so get rust. These are the

22.1

---

[1] *Cf. CP* 1 12. 3.
[2] *Cf. HP* 8 10. 2 (on rust): "And the situation and nature of the field makes no small difference, for fields exposed to the wind and elevated get no rust, or do not get so much, but instead those that lie in hollows and have no wind."
[3] It is also good for the ear to stand well off from the leaf: *cf.* the last sentence of this section.

χρήσιμα,[1] καὶ ὧν ὁ στάχυς ἀπήρτηται πολὺ τῶν φύλλων· ἐν γὰρ τοῖς φύλλοις ἐμμένει μᾶλλον ἡ ὑγρότης, ὥστε πλησίον μὲν ὄντος[2] εὐθὺς ἅπτεται, πόρρω δ' οὐχ[3] ὁμοίως. διὸ καὶ ἡ μὲν Ἀχιλληῒς[4] κριθή, καὶ λευκὴ καὶ μέλαινα, ἐρυσιβώδης, ὀρθοὶ γὰρ οἱ στάχυες, ἡ δ' ἐτεόκριθος ἀσφαλής, ἀπονεύει γάρ. τὰ δὲ προσήνεμα τῶν χωρίων ἧττον ἀπερυσιβοῦνται,[5] διὰ γὰρ τὴν κίνησιν ἀποσείεται καὶ ἀποπίπτει τὸ ὑγρόν. ἐπεὶ καὶ ὅταν ὕσαντος πνεῦμα ἐπιγένηται, καὶ πάλιν ἐπιλαμβάνῃ νύξ, ἧττον· τὸ μὲν γὰρ διέσεισεν, ὁ δ' ἥλιος, εὐθὺς ἐπιγενόμενος, οὐκ ἐποίησεν σῆψιν, ἀλλὰ ἀνεξήρανεν πρότερον. ἡ δ' ἐρυσίβη σαπρότης τις, οὐδὲν δὲ σαπρὸν ἄνευ θερμότητος ἀλλοτρίας. μάλιστα δὲ ἐρυσιβοῦνται σῖτος ταῖς πανσελήνοις, διὰ τὸ[6] καὶ τὴν σελήνην τῇ θερμότητι σήπειν τῆς νυκτός.

---

[1] U : *Iuvat* Gaza : Schneider deletes : ταῦτα διυγραινόμενα Wimmer.
[2] μὲν ὄντος Gaza, Heinsius : μένοντες U : μὲν ὄντες u.
[3] u : ὀχ' U.
[4] U^c : ἀχιλλαῒς U^ac.
[5] ego : γαρ (γὰρ N) ἐρυσιβοῦνται U N : ἐρυσιβοῦνται aP.
[6] Dalecampius : διὸ U.

---

[1] *Cf. HP* 8 4. 2 (of barleys): "... the ears are in some well away from the leaf and in some close to it, as in the so-

DE CAUSIS PLANTARUM III

grains to use, and also those grains where the ear stands well off from the leaves, for the leaf is where the water remains longer, so that when the leaf is close to the ear the disease attacks the ear at once, but when the leaf stands off at a distance the disease is not so catching. This is why the Achilles barley,[1] both the white and the black, is apt to rust, since the ears are erect, whereas the eteocrithos[2] barley is safe, since the ear bends. Fields in the wind are less subject to rust,[3] for the water is shaken off by the buffeting. So when rain is followed by wind and then by night there is less rust, for the wind shakes off some of the water, and the sun does not produce decomposition by coming out at once, but first dries the rest of the water up[4]; and rust is a kind of decomposition, and nothing is decomposed without foreign heat.[5] Grain is most apt to rust at the full moon,[6] since the moon too produces decomposition by its heat during the night.

22.2

---

called Achilles barley"; *HP* 8 10. 2 (of aptness to rust): "... cereals are more apt to rust than pulses, and of cereals barley more than wheat, and of barley some kinds more than others, and most of all (one might say) Achilles barley."

[2] Literally "true-barley"; the name occurs only here.

[3] *Cf. HP* 8 10. 2 (cited in note 2 on *CP* 3 22. 1).

[4] In the milder heat of dawn.

[5] And not the natural heat of the plant.

[6] *Cf. HP* 8 10. 2: "... rust occurs most of all when the moon is full."

THEOPHRASTUS

ἀσθενέστερα δὲ καὶ ἐπικηρότερα[1] πάνθ᾽ ὡς ἁπλῶς εἰπεῖν τὰ λευκὰ τῶν μελάνων καὶ ἐπὶ τῶν φυτῶν ἐστιν καὶ ἐπὶ τῶν ζῴων.

ὅτι μὲν οὖν οἰκεῖα τὰ σπέρματα ταῖς χώραις ληπτέον, φανερὸν ἐκ πολλῶν.

22.3 νοσήματα δὲ γίνεται πᾶσι τοῖς σπέρμασιν (ὡς ἁπλῶς εἰπεῖν) διὰ τὴν ἀσυμμετρίαν τῆς τροφῆς τε καὶ τοῦ περιέχοντος ἀέρος, ὅταν ἡ μὲν πλείων, ἡ δὲ ἐλάττων, ὁ δὲ κάτομβρος ᾖ[2] κατάξηρος ἄγαν ᾖ[3] καὶ μὴ κατὰ καιρὸν ἀνυγραινόμενος τύχῃ. τότε γὰρ καὶ οἱ σκώληκες ἐγγίνονται τοῖς ὤχροις καὶ τοῖς λαθύροις καὶ τοῖς πίσοις, καὶ αἱ κάμπαι τοῖς ἐρεβίνθοις, ὅταν οἱ μὲν ἀναξηρανθῶσιν καὶ[4] μεταξὺ ἐγγίνωνται θερμημερίαι, τῶν δ᾽ ἐρεβίνθων ὅταν ἡ ἄλμη κατακλυσθῇ καὶ ἀναγλυκανθῶσιν· πανταχοῦ γὰρ ἡ φύσις ζῳογονεῖ μιξαμένη πως

---

§ 3 lines 7–8 Athenaeus epitome, ii. 45 (55 E-F): Θεόφραστος δὲ ἱστορεῖ ἐν αἰτίοις φυσικοῖς ... γίνεσθαι δὲ λέγει κάμπας ἐν τοῖς ἐρεβίνθοις ὁ αὐτὸς ἐν τῷ τρίτῳ τῆς αὐτῆς πραγματείας.

[1] Gaza, Wimmer : -όtata ... -όtata U.
[2] u aP : η U : ἢ N.
[3] Gaza (vel), Scaliger : εἰ U.
[4] U : priusquam fuerint resiccata Gaza : <ἀνυγρανθῶσι καὶ πρὶν ἢ> ἀναξηρανθῶσι Schneider : ἀνυγρανθῶσι καὶ Wimmer.

DE CAUSIS PLANTARUM III

Speaking broadly all white varieties are weaker and more delicate both in plants and in animals.[1]

So it is evident from many considerations that we must take grain that is appropriate to the country.

Seed Crops: Diseases (2) Engendering Grubs

Diseases arise in practically all seed crops from the absence of due measure in the food and in the surrounding air—when the food happens to be either too much or too little, and the weather either too rainy or too dry or else precipitation does not occur at the right time. For these are the occasions when the grubs are produced in vetchlings, chicklings and peas, and the caterpillars in chickpeas[2]; when the fluid in the former is made dry[3] and warm days intervene,[4] and when the brine is washed off from the chickpeas and they are made sweet. For everywhere nature generates animals by taking fluid and mixing it in a certain fashion with heat,

22.3

---

[1] *Cf.* Aristotle, *History of Animals*, iii. 21 (523 a 10–11): "... dark women have more wholesome milk than light." We have found no better parallel.

[2] *Cf. HP* 8 10. 5: "Grubs are produced in vetchlings, chicklings and peas when the plants have been soaked and hot days occur, just as the caterpillars in chickpeas."

[3] The wetting provides the superfluous fluid; the drying solidifies it with the help of heat into a grub.

[4] Between the wetting (implied by the drying) and the subsequent drying.

THEOPHRASTUS

τὴν ὑγρότητα τῷ θερμῷ,[1] καθάπερ ὕλην δ' οὖσαν τὴν ὑγρότητα τῷ θερμῷ πρὸς τὴν πέψιν.[2]

22.4 ὃ καὶ ἐπὶ τῶν πυρῶν συμβαίνει κατὰ τοὺς σκώληκας· γίνονται γὰρ ἐν ταῖς ῥίζαις ὅταν νότια πλείω[3] μετὰ τοὺς σπόρους[4] ἐπιγένηται· τότε γάρ, ἀνυγραινομένης τῆς ῥίζης καὶ τοῦ ἀέρος ὄντος θερμοῦ, ζῳοποιεῖ πως ἡ θερμότης συσσήπουσα τὴν ῥίζαν, ὁ δὲ γενόμενος εὐθὺς κατεσθίει· πέφυκε γὰρ ἐκ τῶν αὐτῶν ἑκάστοις ἡ γένεσις καὶ ἡ τροφή.

(ἅτερος δ' ὅταν ἀποχυθῆναι[5] διὰ τὸν αὐχμὸν μὴ δύνηται· <τότε γὰρ> τὸ[6] ἐγκατακλειόμενον ὑγρὸν ὑπὸ τοῦ θερμοῦ σήψεως γινομένης ἐζῳοποί-
22.5 ησεν, εἶθ' ὁμοίως ἡ τροφὴ διὰ τοῦ αὐτοῦ. ταὐτὸ δ'

---

§ 4 lines 1–6 Pliny, *N. H.* 18. 151: nascuntur et vermiculi in radice, cum sementem imbribus secutis inclusit repentinus calor umorem.

[1] ego : τῆι ὑγρότητι τὸ θερμόν U.   [2] U : σῆψιν Heinsius.
[3] u N : νοτια πλείων U : νοτία πλείων aP.
[4] u : τουσπόρους U : τοὺς πόρους N aP.
[5] Schneider (*cf. HP* 8 10. 4) : ἐπιχυθῆναι U.
[6] Gaza (τὸ <γὰρ> Schneider).

---

[1] Heat and water are the source of all life (Aristotle, *On the Generation of Animals*, iii. 11 [762 a 18–21]). Heat concocts the water both in spontaneous generation (*ibid.*, 762

DE CAUSIS PLANTARUM III

the fluid serving as it were as matter for the heat to concoct.[1]

22.4 This also happens in wheat when the grubs are produced.[2] For the grubs are formed in the roots when a good deal of south wind follows the sowing. For then, with the root getting soaked and the weather being warm, the heat in some way generates animals as it helps[3] to decompose the root, and the grub, as soon as produced, proceeds to devour the root, since all things naturally feed on the sources that produced them.[4]

(The other wheat grub is produced when the wheat is unable to head because of the dry weather, since then the fluid shut in by the heat generates animals as decomposition occurs; next these grubs too proceed to get their food from the part that brought them forth.[5] The same process as this 22.5

a 13–15) and in the other kinds (*ibid.*, 762 b 6–16).

[2] For the two kinds of grubs that infest wheat *cf. HP* 8 10. 4: "Wheat is also destroyed by the grubs. Some as soon as the wheat starts to grow devour the roots, others destroy it when the wheat suffers from dry weather and is unable to head; for the grub is then produced in it and eats the haulm as this is played out. It eats the haulm as far as the ear and then perishes when it has consumed all its food."

[3] That is, helps the fluid.

[4] *Cf. CP* 2 9. 5, 6 (of the gall-insects of the fig); *CP* 5 10. 5; 6 4. 4.

[5] Here the haulm.

171

THEOPHRASTUS

ἔοικε τούτῳ[1] καὶ ἐπὶ τῶν μηλεῶν[2] καὶ ὅλως ἐπὶ τῶν δένδρων συμβαίνειν ὅσα σκωληκοῦται διψήσαντα· διὰ γὰρ τὸ ὀλίγον εἶναι τὸ ὕδωρ[3] καὶ μένειν[4] ἐν τῷ δένδρῳ <τὸ ὑγρὸν θερμαινόμενον>,[5] σῆψιν ἐποίησεν, ἐξ ὧν ὁ σκώληξ [ὑγρῷ θερμαινόμενον]. σημεῖον [ἐποίησεν ἐξ ὧν ὁ σκωλιξ][6] <δέ>,[7] ὅταν ἄφθονος ἡ τροφὴ γίνηται· τότε γὰρ καὶ ἡ ἐπιρροὴ πρὸς τὰ ἄνω διαδίδοται, καὶ πλείων[8] οὖσα κατακρατεῖ καὶ οὐ σήπεται.)

παραπλήσιον δὲ τὸ συμβαῖνόν ἐστι καὶ τῇ ἀμπέλῳ· καὶ γὰρ ἐν ταύτῃ τοῖς νοτίοις οἱ[9] ἶπες[10] γίνονται μᾶλλον, ἅτε διυγραινομένης καὶ τοῦ ἀέρος γονεύοντος· εἶτ' εὐθὺς ἐξεσθίουσι[11] τὸ ὁμογενές.

22.6 ὡσαύτως δὲ καὶ ἐπὶ τῶν ἐλαιῶν[12] αἱ κάμπαι, καὶ ἐφ' ὧν ἄλλα ἐγγίνεται ζῷα περὶ τὴν βλάστησιν ἢ τὴν ἄνθησιν ἢ καὶ ὕστερον· πάντα γὰρ ἐκ παραπλησίας αἰτίας ἐστὶ συνιστάμενα. τῇ δ' ἀμ-

---

[1] τούτῳ u : τοῦτο U : τουτὶ N P : a omits.
[2] ego : μηλέων U.
[3] U : ὑγρὸν Wimmer.
[4] u N aP : μενη U.
[5] ego (<διαθερμαινόμενον> Schneider).
[6] σκώληξ ... σκωλιξ U (deletions by Wimmer) : σκώληξ N aP.   [7] Wimmer.
[8] u : πλειον U.   [9] aP : αἰ U (αἱ u N).

appears to occur also in apple-trees and indeed in all trees that get grubs after suffering from drought. For because there is little rain-water and fluid remains in the body of the tree[1] and is heated, it causes decomposition, and the grub results. Proof of this is what happens when the food becomes plentiful, for then the influx is not only passed to the upper parts, but is also greater and masters the heat and undergoes no decomposition.)

What happens to the vine is also close to this, for here too the production of bud-worms is greater when south winds prevail, since the vine is then full of fluid and the weather generative,[2] and the creatures produced proceed at once to eat out what is of the same substance as themselves.

So too in the olive the bend-worms are engendered,[3] and so too with the trees where other creatures are produced at the time of sprouting or flowering or later; for all are formed from a similar cause. This formation happens chiefly in the vine

---

[1] That is, it does not produce foliage or fruit.

[2] That is, warm.

[3] *Cf. HP* 4 14. 9: "At Miletus bend-worms eat the olives at the time of blossom, some the leaves and others the flowers. The two groups differ in kind and lay the trees bare. They occur if there are south winds and clear skies, but if hot weather follows they burst."

---

10 Scaliger (ἴπες u aP) : ἴπε U : ἴπαι N.
11 aP c? : -εθί- U N P ac(?).   12 u aP : -αίων U N.

THEOPHRASTUS

πέλῳ μάλιστα τοῦτο συμβαίνει διότι φύσει ὑγρόν ἐστιν, καὶ ἡ ὑγρότης αὐτῆς ἄχυλος καὶ ὑδατώδης · εὐπαθεστάτη γὰρ ἡ τοιαύτη. ἐνιαχοῦ δὲ οὐ γίνονται τὸ ὅλον ἶπες,[1] ὅταν εὔπνους τε καὶ μὴ ἔνυγρος μηδ᾿[2] εὔτροφος ὁ τόπος ὑπάρχῃ.

καὶ περὶ μὲν νοσημάτων ἐν τοῖς ὕστερον ἐπὶ πλέον ῥητέον.

23.1 σπείρειν[3] δὲ κελεύουσιν οἱ μὲν πρὸ Πλειάδος · ξηρὰν γὰρ ἄνικμόν τε οὖσαν διαφυλάττειν τὸ σπέρμα τὴν γῆν · οἱ δ᾿ ἅμα Πλειάσι δυομέναις, ὥσπερ καὶ Κλείδημος ·[4] ἐπιγίνεσθαι γὰρ ὕδατα, καὶ πολλά, τῇ ἑβδόμῃ μετὰ τὴν δύσιν. ἄριστον δὲ ἴσως καὶ ἀσφαλέστατον εἰς ὀργῶσαν ἐπειδὰν ἐμβληθῇ,[5] διευλαβούμενον ὅπως μήτ᾿ ἐμβληθῇ πηλῷ μήτ᾿ <εἰς>[6] ἡμιβραχῇ καὶ ἡμίειλον,[7] ἣν δὴ καλοῦσί τινες ἀμφίεργον · ἡ μὲν γὰρ ὑγρὰ καὶ πηλώδης διαχεῖ καὶ ἐκγαλακτοῖ, καὶ ἐὰν ξηρανθῇ

§ 1 line 4 and § 2 lines 3–9 Cleidemus, Fragment 5, Diels-Kranz, *Die Fragmente der Vorsokratiker*, vol. ii⁸, p. 50.

[1] Schneider : ἶπες U.   [2] aP : ἢ δ᾿ U : ἢ δ᾿ u : ἢ δ᾿ N.
[3] u : ἐπιρρεῖν U N aP.   [4] aP : κλήδημος U : κλήδιμος N.
[5] ἐπειδ᾿ ἂν ἐμβληθῇ U : ἐμβάλλειν (*semen committere* Gaza) Schneider : τὴν γῆν ἐμβληθῆναι Wimmer.
[6] Wimmer.   [7] U : ἡμιβραχεῖ καὶ ἡμιή<λῳ> Heinsius.

because it is naturally fluid, and its fluid is without savour and watery, for fluid of this character is the most easily affected. But in some places no bud-worms are produced at all, when the region is a well-ventilated one and not well-watered or a provider of abundant food.

But diseases will be discussed more fully later.[1]

Seed Crops: The Seasons of Sowing

Some authorities tell us to sow before the rising of the Pleiades, for when the soil is dry and lacks moisture it preserves the seed. Others tell us to sow at the setting of the Pleiades, like Cleidemus, for rains, and heavy ones, come on the seventh day after the setting. But perhaps the best and safest maxim is to see to it that when the seed is sown the earth shall be in heat, taking care that the sowing shall not have been made in mud or in soil half-wet and half-sunned, which some call *amphiergos*,[2] since wet and muddy ground dissolves the seeds and makes them milky,[3] and when it dries seals them in

23.1

---

[1] *CP* 5 8. 1–5 10. 5.

[2] "Worked both ways"; the word is not elsewhere attested.

[3] *Cf. HP* 8 6. 1: "It is considered that sowing in the early season (*sc.* at the setting of the Pleiades) is on the whole better, and that the worst sowing is that in half-wet ground, for the seeds perish and turn milky."

THEOPHRASTUS

συναλείφει καὶ οὐ διαδίδωσιν, ἡ δὲ ἡμιβραχὴς[1] κατασήπει· τοσοῦτον γὰρ ἔχει τὸ θερμὸν καὶ τὸ ὑγρὸν ὥστε κινῆσαι μέν, μὴ ἐκβιάσασθαι δὲ μηδὲ

23.2 ἐκβλαστεῖν. ᾗ[2] καὶ χείριστος οὗτος[3] τῶν σπόρων. ὁ γὰρ ἐν ξηρᾷ σπόρος[4] γινόμενος καθ' ὥραν ἔτους οὐκ ἀπόλλυσιν τὸ σπέρμα· τοὺς δὲ περὶ τροπὰς σπόρους Κλείδημός φησιν ἐπισφαλεῖς εἶναι· διερὰν γὰρ οὖσαν καὶ βαρεῖαν τὴν γῆν ἀτμιδώδη γίνεσθαι, καὶ ἐοικέναι ἐρίοις κακῶς ἐξαμμένοις·[5] ἔτι δ' οὐ δύνασθαι τὰς ἀτμίδας ἕλκειν, οὐδὲ διαπέμπειν, ἅτε θερμὸν οὐκ ἔχουσαν ἱκανόν, καὶ ἐπαλείφιν ἔλαττον. ἀλλὰ τὰ ὄψια δεῖ σπείρειν ὑπὸ τροπάς.

23.3 συμφέρει δὲ σπειρομένοις καὶ σπαρεῖσιν εὐδίας ἐπιγενέσθαι[6] πλείους ἡμέρας, ὅπως διαβλάστω-

§ 2 lines 4–6 *Geoponica*, iii. 1. 7: τῷ αὐτῷ μηνὶ (January) σπείρειν οὐ χρή, ἀραιὰ γὰρ καὶ βαρεῖα οὖσα ἡ γῆ ἀτμώδης γίνεται, καὶ ἔοικεν ἐρίοις ἐξαμμένοις κακῶς.

[1] N a (-ὺς P) : ἡμιβρεχῆς U (ἡμιβρεχὴς u).
[2] Gaza (*quamobrem* : διὸ Heinsius : Wimmer deletes) : ἢ U (ἢ u N) : εἰ aP.
[3] u : οὕτως U.
[4] Heinsius after Gaza : σπορᾷ U.
[5] Schneider : εξεσμενοις U. [6] U N : ἐπιγίνεσθαι aP.

DE CAUSIS PLANTARUM III

and allows no passage, and on the other hand half-wet ground decomposes them, for it has just enough heat and fluid to set them growing but not enough to let the plant force its way through or come up. This makes it the worst of the sowings.[1] For sowing on dry soil in the spring does not destroy the seed, whereas sowing at the winter solstice is called risky by Cleidemus, who says that the soil, which is soaked and heavy, gets filled with vapour and resembles badly carded wool[2]; and that it furthermore is unable to absorb or diffuse the pockets of vapour since it does not possess enough heat, and does not form a crust over the seed to the same extent. Late grains however must be sown at the winter solstice.[3]

Seed Crops:
The Best Weather During and After Sowing

During sowing and just after it several days of fair weather are good, to let the seeds sprout[4]; after

---

[1] *Cf. HP* 8 6. 1, quoted in preceding note.

[2] It is lumpy, just as poorly carded wool has snarls in it.

[3] This is the late sowing of *HP* 8 1. 2, "at the beginning of spring after the winter solstice." The "late" grains are those that are sown in the "late" sowing, and are listed in *HP* 8 1. 4.

[4] *Cf. HP* 8 1. 5: wheat and barley come up on about the seventh day after sowing; most legumes on the fourth or fifth.

THEOPHRASTUS

σιν,[1] μετὰ δὲ ταῦτα ψύχη βόρεια καὶ μέτρια [λεπτὰ][2] χειμῶνος, ἰσχυρῶς ἐρριζωμένων ἤδη καὶ ταρρουμένων ·[3] ἀνθέξουσι γὰρ μᾶλλον, ἐπεὶ πιλούμενά[4] γε βελτίω καὶ ἐγκαρπότερα, τῆς γὰρ ῥίζης ἰσχυούσης, ἥ τε τροφὴ πλείων καὶ ἡ αὔξησις μετὰ ταῦτα θάττων ·[5] τὰ γὰρ εὐθὺς ἀνατρέχοντα λεπτὰ καὶ ἀσθενῆ γίνεται, καὶ ἅμα, πρὸ τῆς[6] ὥρας κυϊσκόμενα καὶ ἀποχεόμενα, φθείρεται. τὸ γὰρ ὅλον ἡ πάρωρος εὐτροφία σφαλερά (διὸ καὶ ἐπινέμουσιν,[7] οἱ δὲ κείρουσιν).

23.4 ἀγαθὸν δὲ καὶ ἡ χιών, ὅτι ἀναζυμοῖ[8] καὶ μανοῖ τὴν γῆν · καὶ τροφήν τε παρέχει, καὶ ἐγκατακλείουσα τὸ θερμὸν αὔξει[9] τε καὶ ἰσχύειν ποιεῖ τὴν ῥίζαν. ᾗ[10] καὶ τὸ ἀπορούμενον οὐκ ἀφανές, διὰ τί <καὶ>[11] αἱ ψυχειναὶ χῶραι καὶ αἱ θερμαὶ σιτοφόροι (καθάπερ ἡ Θρᾴκη καὶ ὁ Πόντος,[12] καὶ ἡ Λιβύη καὶ ἡ Αἴγυπτος) · πρὸς γὰρ τοῖς ἄλλοις ποιεῖ τι ὁ χειμὼν καὶ τὰ καύματα (καθάπερ ἐν ταῖς κατερ-

---

[1] ego : διαβλαστῶσιν (-σι U^r) U^{ar}.
[2] Gaza (a variant of μέτρια?).
[3] καὶ τ. Wimmer : καταρρουμένων U.
[4] ἐπεὶ π. (a omits -α) P : ἐπιπηλουμενά U : ἐπιπιλούμενά u (-α N).   [5] a : θᾶττον U N P.   [6] u : πρωτῆς U.
[7] ego : ἐπιτέμνουσιν U.   [8] Schneider : ὅτ' ἂν ζυμοῖ U.
[9] Schneider : -ειν U.   [10] ᾗ aP : ἡ U : ἢ u N.
[11] ego.   [12] Gaza : ποταμὸς U.

DE CAUSIS PLANTARUM III

that, in winter, northerly and moderate cold weather, when the grain has already become strongly rooted and tillered, for it will then better resist the cold (indeed when the cold compresses it the grain is finer and yields a bigger crop). For when the root is strong there is more food, and the growth, which comes later, is more rapid; for plants that spring up at once turn out to be thin and weak, and then too, since they begin to be pregnant with their crop and to head before the spring season, are killed by the weather. In a word good feeding at the wrong season is hazardous (which is why people graze the plants and others cut them down).[1]

Snow too is beneficial, because it ferments[2] the soil and gives it an open texture; again it furnishes food and by shutting in the heat makes the root grow and gives it strength. This shows the solution of the puzzle: why is it that both cold countries (as Thrace and Pontus)[3] and hot (as Libya[4] and Egypt) are producers of grain? For besides their other effects the winter and the hot weather (as we said[5]

23.4

---

[1] *Cf. HP* 8 7. 4: "In fertile countries, to prevent running to leaf, people graze the grain and cut it down, as in Thessaly.... In Babylonia ... they cut it down twice and the third time turn the sheep loose in it ..."

[2] *Cf.* Theophrastus, *On Fire*, 2. 18.

[3] The Crimea and the Ukraine.

[4] Carthage.

[5] *CP* 3 20. 7.

γασίαις ἐλέχθη τῆς γῆς). οὐδὲν δὲ ἔλαττον, ἀλλὰ πάντων μέγιστον, ὁ ἀὴρ ὁ περιέχων αὐτάς, πρὸς εὐκαιρίας ὑδάτων καὶ[1] χειμώνων καὶ πνευμάτων. ὁποῖα γὰρ ἂν ᾖ ταῦτα, καὶ τὰ σπέρματα οὕτως ἐκτελεῖται, ὃ καὶ ἡ παροιμία καλῶς "ἔτος φέρει, οὔτι ἄρουρα."

23.5 μέγα δὲ καὶ ἡ θέσις τῆς χώρας ἡ[2] πρὸς τὰ πνεύματα καὶ τὸν ἥλιον, ὥσπερ καὶ ἐπὶ τῶν δένδρων ἐλέχθη· πολλαὶ γὰρ οὖσαι λεπταὶ καὶ φαῦλοι τελεσφόροι γίνονται διὰ τὸ πρὸς ταῦτα κεῖσθαι καλῶς. ὡς δ' ἁπλῶς εἰπεῖν ἡ ἀγαθὴ χώρα καὶ χειμώνων δεῖται πολλῶν· ἂν[3] γὰρ εὐδίαι καὶ τὰ νότια ἐνισχύωσιν λοχαίους[4] ποιοῦσιν ὡς ἐπὶ τὸ πολὺ καὶ ἐρυσιβώδεις, ὁ δὲ χειμών, πιλώσας καὶ καρκινώσας τὰς ῥίζας, σύμμετρον εἰς τὸ ἔαρ ποιεῖ

---

[1] U : καὶ βία N : καὶ βίας aP. (The variants are due to misreading U^m.) [2] u : ἢ U. [3] a : αἱ (αἲ U) u N P.
[4] Schneider : λεχεους U : λεχαίους u.

---

[1] *Cf. HP* 8 7. 6: "The greatest contribution to growth and feeding is made by the blend of qualities in the air, and in general by the way the year turns out; for when rain, clear weather and cold occur at the right time all the seeds bear well and yield abundant crops, even if they are in sandy and lean land, which is why people do not put the matter

in discussing the working of the ground) do something to the soil; but the air surrounding these countries does something that is in no way less, but most important of all, in bringing about rains and cold and winds at the right time; for as these go, so goes the final development of the seed, and the proverb also puts this well: "The harvest is the year's and not the field's."[1]

Seed Crops: The Lie of the Land

The lie of the land to wind and sun is also of great importance (as we said[2] in discussing trees), for many a lean and poor soil brings the seeds to fruition because it is well situated with regard to these.[3] Good land, to put it broadly, also requires plenty of storm and cold, since if fair weather and south winds prevail they for the most part cause the grain to lodge and get rust. Cold weather on the other hand compresses the plants and tillers the roots, and so gives them the right size when spring

23.5

---

badly in the proverb: 'The harvest is the year's and not the field's.'"

[2] *CP* 2 3. 1–3, 3 6. 9 (for wind), 2 4. 8 (for sun).

[3] *Cf. HP* 8 7. 6: "Countries too also differ greatly, not only by being fat or lean and rainy or dry, but also by the surrounding air and the winds; for some which are lean and poor bring the grain to maturity because they are well situated with regard to the winds from the sea."

THEOPHRASTUS

τὸ μέγεθος, ὅλως δὲ καὶ τὴν χώραν, ἂν εἰργασμένην[1] λάβῃ, μανοῖ. διὰ ταύτην δὲ τὴν αἰτίαν καὶ εἰς τὸ ἔαρ αἱ αὐξήσεις ταχεῖαι γίνονται καὶ <αἱ>[2] ἐκφύσεις καὶ <αἱ>[2] τελειώσεις, ὥστε δοκεῖν μὴ κατὰ λόγον ἐπιλείπεσθαι τῶν ἐν ταῖς ἀλεειναῖς, ἀλλὰ καὶ προτερεῖν, ὥσπερ καὶ τὰ περὶ Ἑλλήσποντον· ἰσχυρᾶς <γὰρ>[3] τῆς ὁρμῆς γενομένης διὰ τὸν ἀθροισμόν, καὶ τοῦ ἀέρος ὑπηρετοῦντος,[4] ταχεῖαι καὶ <αἱ>[5] ἐπιδόσεις καὶ αἱ τελειώσεις γίνονται.

ταῦτα μὲν οὖν ἐν ταῖς τῆς χώρας[6] διαφοραῖς ἐστι.

24.1 τὰ δὲ σπέρματα, καθάπερ ἐπὶ τῶν γύρων,[7] ἐκ τῶν ὁμοίων καὶ[8] τῶν χειρόνων ληπτέον, ὅπως δὴ[9] μηδεμία μεταβολὴ <καὶ>[10] ἐπὶ τὸ βέλτιον γίνηται. καίτοι φασὶν ἐκ τῆς ἀγαθῆς ἰσχυρότερα εἶναι, διὸ καὶ ἐπὶ δύο ἔτη διαμένειν τὴν δύναμιν. εἰ δὲ τοῦτο ἀληθές, ὅμοιον ἂν εἴη καὶ τὸ ἐπὶ τῶν

---

[1] ἂν εἰργ. u aP : ἀνειργασμένην U N.
[2] Schneider.
[3] Gaza, Schneider.   [4] a : ὑπηροῦντος U (ὑ- N P).
[5] aP.   [6] τῆς χώρας u aP : τῆς χώρας U : χώραις N.
[7] ego (δένδρων Schneider) : πυρῶν U.
[8] U : <ἢ> καὶ Schneider : ἢ Wimmer.
[9] U : ἢ Schneider.   [10] ego (<ἢ> Schneider).

DE CAUSIS PLANTARUM III

comes round; and as for the land, if it finds the land already tilled, cold weather generally gives it an open texture. For this reason when spring comes round growth is rapid, and so too is heading and maturing, so that it is considered that grain in cold country is not as far behind grain in warm country as would correspond to the later spring, but is in fact in advance of it, as with Hellespontine grain.[1] For when the impulse to grow has become strong because of the accumulated food, and the weather furthers it, the grain both develops and ripens rapidly.

These matters, then, depend on the differences of country.

Seed Crops: Choice and Treatment of the Seed

24.1 The seed (as with the planting in the holes)[2] should be taken from land as good or land that is inferior, so that there will be no change or a change for the better. Yet we are told that seed from good land is stronger (which is why its power remains for two years).[3] If this is true, the rule would be like

---

[1] *Cf. HP* 8 2. 10: "For the grain at Athens is some thirty days (or a little more) earlier than the grain at the Hellespont. Now if it is also sown earlier, the shift would be of the dates; but if it is sown at the same time, the period would evidently be longer." [2] *CP* 3 5. 2.

[3] Foreign grain is assimilated in the third year: *CP* 1 9. 3, 2 13. 3, 4 1. 6.

THEOPHRASTUS

φυτευμάτων, τὸ[1] τὰ κάλλιστα καὶ ἰσχυρότατα λαμβάνειν, ὥστ' ἀμφοτέρως ἂν ἔχοι λόγον.

24.2 χρὴ δὲ καὶ ταῖς τοῦ ἀέρος ἀνωμαλίαις προσέχειν· οὔτε γὰρ τὰ ἐκ τῶν εὐείλων καὶ πρωΐων[2] εἰς τοὺς ὀψίους καὶ δυσχειμέρους, οὔτε τὰ ἐκ τούτων εἰς ἐκείνους ἁρμόττειν·[3] τὰ μὲν γὰρ προτερεῖν δοκεῖ, τὰ δ' ὑστερεῖν, ὥστε τὰ μὲν ὑπὸ χειμῶνος, τὰ δ' ὑπ' αὐχμῶν ἀπόλλυται καὶ ἀνυδρίας (ὄψια γὰρ ὄντα βραδέως τε κυΐσκεται καὶ ἀποτίκτει, διὸ καὶ ὑστερεῖ). ταὐτὸ δὲ τοῦτο συμβαίνει καὶ ἐπὶ τῶν σπερμάτων ὅσαπερ[4] εὐβλαστεῖ καὶ

24.3 ταχὺ παραγίνεται, καθάπερ τὰ τρίμηνα, ἃ[5] σπαρέντα[6] πρώϊα προτερεῖ καιρῶν,[7] εἰ μὴ[8] τοιοῦτόν ἐστιν ὡς προλαβεῖν[9] τοῦ χειμῶνος[10] πρὸς τὴν ῥίζωσιν, ὥσπερ καὶ τὸν κύαμον διὰ τοῦτο πρωϊσπο-

---

[1] u : τῷ U.

[2] U : πρωΐνων Wimmer.

[3] u N a (ἀ- U) : -ει P.

[4] Wimmer (ὅσα Schneider) : ὅσα γὰρ U.

[5] ego (*trimenstria: haec* Gaza : τριμηναῖα Lobeck [Phrynichus, p. 549]) : τριμηναῖα U^c (from -αία).

[6] σπαρέντα ⟨γὰρ⟩ Schneider.

[7] U : καιρὸν u.

[8] μή ⟨τι⟩ Gaza, Schneider.

[9] ego (*sit: ut ... anticipare debeat* Gaza) : ἐστιν ὃ πρ. ⟨δεῖ⟩ Schneider : ὥστε πρ. Wimmer) : ἐστιν ὃ πρ. U.

[10] U : τοὺς χειμῶνας Schneider (*cf. CP* 2 12. 5).

DE CAUSIS PLANTARUM III

that about cuttings: to take the finest and strongest,[1] so that either choice would be reasonable.

We must also attend to differences in the air of the localities concerned. So we are advised that it is neither suitable to sow seed from a sunny and early-producing region in one that produces late and has a harsh winter, nor to sow seed taken from the latter region in the former. For seeds taken from the former are considered to come out too early, those from the latter too late, and so the former are killed by cold, the latter by dry weather and lack of rain, since being late sprouters they are slow to head and bear, which makes them too late.[2] The same happens with those grains that sprout well and come up rapidly (like the three-months varieties).[3] These when sown early[4] develop too soon, unless the plant needs to get rooted before winter sets in, just as they sow bean early for this

24.2

24.3

---

[1] *CP* 3 5. 1.

[2] *Cf. HP* 8 8. 1: "Different kinds of seeds also suit the natures of different countries ... It is good to make the change from warm localities to those slightly less warm, and so with cold localities. Seeds taken from localities with harsh winters are in early producing localities late in heading, and so are destroyed by rainless weather (unless rain falls late and saves them)."

[3] For three-months wheat and barley (sown in spring) *cf. CP* 4 11. 1.

[4] At the setting of the Pleiades.

THEOPHRASTUS

ροῦσιν, καὶ ἅμα διὰ τὸ φιλεῖν ἀνθοῦντα βρέχεσθαι, καὶ ἀνανθεῖν[1] πολὺν χρόνον · ὀψὲ γὰρ σπειρόμενος οὐκ ἂν βρέχοιτο. τὸ δὲ ὕδωρ ἐν τῷ ἀνθεῖν τούτῳ[2] μὲν συμφέρειν[3] διὰ τὴν μανότητα, τῷ σίτῳ δὲ ἀσύμφορον διὰ τὴν λεχθεῖσαν πρότερον αἰτίαν · τοῖς δὲ ἄλλοις χεδροποῖς ἀβλαβὲς πλὴν ἐρεβίνθῳ, τοῦτον δὲ ἀπολλύει ·[4] καταπλυθείσης γὰρ τῆς ἅλμης, ὥσπερ συμφύτου τινὸς στερόμενος, σφακελίζει τε καὶ ὑπὸ καμπῶν κατεσθίεται (ζῳογονεῖται γάρ, ὥσπερ εἴρηται).

24.4 βραχέντος δὲ οἴνῳ τοῦ σπέρματος ἧττον δοκεῖ νοσεῖν, ὅπερ οὐκ ἄλογον, ἔχοντί[5] τινα δριμύτητα, διατηροῦσι γὰρ αὐταί. τὸ γὰρ ὅλον, ἐάν τις ἀλλοι-

§ 4 line 1 Pliny, N. H. 18. 157 (of seed crops): vino ante semina perfusa minus aegrotare existimant.

line 1 Geoponica, ii 18. 6: Ἀπουλήϊος δέ φησι τὰ οἴνῳ ἐπιρρανθέντα σπέρματα ἔλαττον νοσήσειν.

[1] U : ἀπανθεῖν Wimmer.
[2] τούτῳ u aP : τοῦτο U N.
[3] U : συμφέρει u.
[4] Schneider : ἀπολύει U.
[5] ego : -τα U.

[1] Cf. HP 8 1. 3 (of seed crops sown early): "Of legumes bean and birds' pease are (one may say) the ones that are most of all sown early, for on account of their weakness these like to anticipate the cold weather in getting rooted."

DE CAUSIS PLANTARUM III

reason[1] and then too because it likes rain when it flowers and puts out flowers over a considerable time,[2] for if sown late it would get no rain. Rain at flowering time (we are told) is good for bean because of its open texture, but it is bad for cereals for the reason mentioned earlier.[3] It is harmless to the other legumes with the exception of chickpea, which it destroys, since chickpea, as if deprived of something belonging to its nature when the brine is washed off, gets necrosis and is devoured by bend-worms[4] (animals being produced in it as we mentioned).[5]

If the seed has been wet with wine the plant is considered to suffer less from disease. This is not unreasonable, since the wine has a certain pungency, and pungencies are preservative. For in

24.4

---

[2] *Cf. HP* 8 2. 5: "But in legumes flowering lasts a long time, longer in vetch and chickpea than in the others, but longest of all of these and with the greatest difference from other seed crops in bean, for they say that it blooms for forty days..."; *HP* 8 6. 5: "Bean is fondest of water when it blooms, which is why people are unwilling to sow it late, because (as we said) it blooms for a long time..."

[3] *CP* 2 2. 2; *cf. HP* 8 6. 5 (of rain): "... but it is harmful to wheat, barley and the other cereals when they are in bloom, since it destroys the flowers..."

[4] *Cf. CP* 3 22. 3; *HP* 8 6. 5 (on rain): "For pulse it is harmless, except for chickpea, for when their brine is washed off these are destroyed by necrosis and by being devoured by bend-worms..."   [5] *CP* 3 22. 3.

THEOPHRASTUS

ὥσας καταβάλῃ τὸ σπέρμα (καθάπερ πρότερον ἐλέχθη) μεταβάλλειν καὶ <τὰ>[1] φυτὰ καὶ τοὺς καρποὺς οὐκ ἄλογον · οἷα γὰρ ἡ ἀρχή, τοιαῦτα καὶ τὰ ἀπὸ τῆς ἀρχῆς. ἔτι[2] δ' ὁρῶμεν ὅτι καὶ ταῖς τροφαῖς ἀλλοιούμενα μεταβάλλει καὶ ὅλα γένη τῶν δένδρων (ὥσπερ ἐξ ἀγρίων ἥμερα γινόμενα καὶ ἐκ μελάνων λευκά), καὶ κατὰ τοὺς καρποὺς ὁμοίως · ὥστε καὶ πρὶν πεσεῖν, αὐτὸ ποιόν τι γινόμενον,[3] οὐκ ἄτοπον ἀλλοιοῦν καὶ τὴν ῥίζαν εὐθύς, καὶ τὸν καυλόν, καὶ ἔσχατον τὸν καρπόν.

ἀπηνθηκόσι[4] δ' οὐ συμφέρει κατὰ μικρὸν ἐφύειν,[5] οὐδ' ἡλίους ἐπιλαμβάνειν, ἀλλὰ ψύχη γίνεσθαι καὶ ἐπινεφεῖν,[6] ὅπως μὴ ἐπικαθήμενον τὸ ὕδωρ ἐπιλαβὼν ὁ ἥλιος ἐρυσιβώσῃ[7] σῆψιν[8] ποιήσας · διὸ καὶ αἱ δροσοβόλοι χῶραι καὶ ἔγκοιλοι καὶ ἀπνεύματοι μάλιστα ἐρυσιβώδεις.[9]

---

[1] v^{ac}, Schneider.
[2] u : εἴ τι U.
[3] aP : -αι U N.
[4] Schneider : -κὸς U.
[5] Schneider (*irrorari* Gaza) : ἐκφύειν U.
[6] Schneider : ἐπινεφειν U (-νέφειν N aP : -νίφειν u).
[7] N aP (-ώσῃ u) : ἐρυσιβῶσι U.
[8] Scaliger : ψηλὴν U : ψήλην u.
[9] subscription: θεοφράστου περι φυτῶν ἱστοριας τὸ γ̄ U.

DE CAUSIS PLANTARUM III

general, if one alters the seed before sowing (as we said before),[1] it is not unreasonable that both the plant and the crop should change, for as goes the beginning, so goes what comes from the beginning. Again we observe that not only the entire kind of a tree is altered and changed by the food (as when the tree becomes a cultivated one instead of wild[2] and white instead of black),[3] but that this happens similarly with seed crops, so that it is not strange that when the seed before sowing acquires a certain character itself, it should also change the character of the root to begin with, then of the stem, and last of the fruit.

After the flower is shed it is not advantageous for short rains to follow, succeeded by sunshine, but rather for there to be cold weather[4] and an overcast sky, so that the sun may not catch the rainwater while it is still on the plant and cause rust by bringing about decomposition. This is why countries with a fall of dew and that lie in a hollow and have no wind are especially liable to rust.[5]

---

[1] *CP* 3 9. 4.
[2] *Cf. CP* 2 14. 1.
[3] *Cf. CP* 2 13. 2.
[4] Cold has a drying effect.
[5] *Cf. HP* 8 10. 2, cited in note 2 on *CP* 3 22. 1.

Δ[1]

1.1 ἐπειδὴ πλείους αἱ γενέσεις τῶν φυτῶν, τούτων δὲ κοινοτάτη πᾶσιν ἡ ἀπὸ σπέρματος (ἢ ὅσα ἔχει σπέρμα καὶ καρπόν), ἀπορήσειεν ἄν τις διὰ τί ποτε οὐκ ἀνὰ λόγον[2] ἡ ἰσχὺς ἑκάστοις ἐστὶ τῶν σπερμάτων, ἀλλ' ἀσθενέστερα[3] <τὰ>[4] τῶν ἰσχυροτέρων, οἷον τὰ τῶν δένδρων τῶν ἐπετείων. ἀμφοτέρως γὰρ φαίνεται τὸ ὁμολογούμενον· καὶ εἰ ἀπὸ ἀσθενεστέρων σπερμάτων ἰσχυρότερα <τὰ>[5] γινόμενα, τὰ γὰρ σπέρματα γεννᾷ τὰ δένδρα. τῆς δὲ ἀσθενείας καὶ τῆς ἰσχύος ἐκεῖνα σημεῖα· τὰ μὲν γὰρ ἐπέτεια, ὅταν σπαρῇ, διαμένει τὰ αὐτὰ καὶ ἐξομοιοῖ τοὺς καρπούς, τὰ δὲ τῶν δένδρων μεταβάλλει καὶ χείρω ποιεῖ· καὶ τὰ μὲν ταχεῖαν

[1] τὸ δ̄ U^m.   [2] Wimmer : ἀνάλογον U.
[3] u N : -εστερα U : -εστέρα aP.
[4] ego (quae a validioribus prodeant Gaza).
[5] Schneider.

BOOK IV

A Difficulty: The Weaker Seeds Produce and Are Produced by the Stronger Plants

Since of the several modes of generation in plants the one most common to all (or common to all that have seed and fruit) is generation from seed,[1] one might raise a difficulty: why the strength of the seed does not answer to the strength of the various plants, but the weaker seeds come from the stronger plants, to wit, the seeds of trees are weaker than those of annuals. (For put it either way, and the fact is admitted: you can also say that the weaker seed produces the stronger plant, since the seed generates the tree.) Proof of this weakness and strength is the following: when annuals are sown, the plant produced is the same as its parent and produces the fruit that its parent produced, whereas the seeds of trees change the character of the fruit produced and make an inferior tree[2]; again, the seeds of annuals grow up and mature quickly,

1.1

---

[1] Cf. CP 1 1.1.
[2] Cf. CP 1 9.1; HP 2 2.4–6.

THEOPHRASTUS

ἀποδίδωσιν τὴν τελείωσιν, τὰ δὲ χρόνιον καὶ βραδεῖαν. καὶ ἀνάπαλιν δὲ λαμβανομένων· ἰσχυρότερα γὰρ ὄντα τὰ δένδρα βραδύτερον ἐκπέττει τοὺς καρπούς.

ἔτι δὲ τὰ μὲν ἐν κελύφεσι δερματικοῖς καὶ ξυλώδεσιν, τὰ δ' αὐτὰ πάλιν ἐν περικαρπίοις σαρκώδεσιν (οἷον τὸ τῆς ἐλάας καὶ τὸ τῆς κοκκυμηλέας, καὶ τὰ τῆς ἀμπέλου καὶ ἀπίου καὶ μηλέας), ἔνια δὲ ἐν ξυλώδεσιν ἅμα καὶ σαρκώδεσιν καὶ ἄλλως,[1] ὅσα κάρυον ἐντὸς ἔχει τὴν σάρκα.[2] τὰ δὲ τοῦ σίτου γυμνά, καὶ μάλιστα τοῦ πυροῦ καὶ τῆς κριθῆς, εἰ δὲ μή, χιτῶσί γε περιέχεται λεπτοῖς. αἰεὶ δὲ τὸ ἀσθενέστερον ἡ φύσις εἰς πλείω τίθεται φυλακήν.

αἱ μὲν οὖν ἀπορίαι σχεδὸν αἱ αὐταί.

δεῖ δὲ λαβεῖν πρὸς αὐτὰς ἀρχὴν τήνδε· τὸ

[1] ego : ὅλως U.   [2] ego : τῆς σαρκὸς U.

[1] That is, trees take longer to produce mature seed. *Cf.* Aristotle, *On the Generation of Animals*, i. 18 (724 b 19–20): "Seed and fruit differ in being prior or posterior: a fruit by being from something else, a seed by something else's being from it, since both are the same thing."

[2] *Cf. HP* 1 11. 3: "No tree seed is naked, but is either enclosed in flesh or in a hull, some of the seeds in leathery hulls, as acorn and sweet chestnut, some of the seeds in

DE CAUSIS PLANTARUM IV

whereas the maturing of trees from seed is late and slow. So too when we take plant and seed the other way: trees, the stronger plant, take longer to complete the concoction of their fruit.[1]

Again, the seeds of trees are (1) in some cases enclosed in a leathery or woody cover, the cover itself in turn being enclosed in a fleshy pericarpion (for example the seed of olive and plum and the seeds of vine, pear and apple), and (2) in other cases the seed has the combination of woody and fleshy enclosure in a different arrangement, when the seed has the flesh inside in the form of a nut.[2] Whereas the seeds of cereals are naked, and most of all the seeds of wheat and barley (or if not naked, the seeds are enclosed only in thin coats). But the nature of a plant always surrounds the weaker thing with the greater protection.

So the difficulties are (one may say) the same.[3]

1.2

The Solution

To meet these difficulties we must start from the

1.3

woody hulls, as almond and filbert ... Of the seeds themselves (*sc.* as opposed to the hull), in some plants they are flesh from the start (*sc.* as you pass from outside inward), as with nut-like and acorn-like seeds, but in others the fleshy part is contained in a stone, as with the olive, bay and others."

[3] Whether one takes the seed as product or source of the plant.

σπέρμα μὴ μόνον ἔχειν δύναμιν [τοῦ] τῷ[1] ποιεῖν ἀλλὰ καὶ τῷ[2] πάσχειν, ὥσπερ[3] καὶ κατὰ πάντων τῶν τῆς φύσεως ἀληθές· καὶ κατ' αὐτὴν δὲ τὴν γένεσιν ἀμφοῖν εἶναι τὴν ἐνέργειαν, καὶ τρόπον τινὰ οὐχ ἧττον τῷ[4] πάσχειν, τότε γὰρ καὶ ἡ ἐν τοῖς σπέρμασιν κινεῖται δύναμις οἷον διαθερμαινομένων. διὸ καὶ οὐ πάντα βλαστάνει κατὰ τὴν αὐτὴν αἰτίαν,[5] ἀλλ' ὅταν ἡ αἰτία[6] τοῦ ἀέρος κατάσχῃ κρᾶσις.

ὑποκειμένου δὲ τούτου, φανερὸν ὡς οὐκ ἰσχύος ἀλλ' ἀσθενείας μᾶλλον ἡ ταχυβλαστία· τὸ γὰρ ἀσθενὲς εὐπαθέστερον (διὸ καὶ τὰ [εὐπαθέστερα][7] ἐπέτεια[8] ταχυβλαστότερα, καὶ ἅμα τὸ γυμνὸν[9] αὐτοῖς[10] ὥσπερ ἕτερον εἰς τὸ παθεῖν).

1.4 ὁ δ' αὐτὸς λόγος καὶ περὶ τῆς τελειώσεως· τὰ <γὰρ>[11] ἀσθενέστερα ῥᾷον ἐκτελειώσασθαι καὶ τῷ περιέχοντι καὶ ταῖς ἐν αὐτοῖς[12] ἀρχαῖς. ὃ καὶ περ

---

[1] ego : τοῦτο U : τοῦ u.
[2] ego : τοῦ U.
[3] U : ὅπερ u.
[4] U : τοῦ Gaza, Schneider.
[5] U : ὥραν Schneider.
[6] U : οἰκεία Gaza (aptus), Itali.
[7] ego.
[8] u (N aP omit διὸ ... ἕτερον by homoioteleuton): ἐπέτεια U. [9] ego : κοινὸν U.

DE CAUSIS PLANTARUM IV

following principle: that the seed has not only a capacity by way of action, but also a capacity by way of being acted upon, as is also true of all performances of the nature of a plant; and that moreover in the very generation of a plant the operation is realized in both ways, and no less (in a fashion) by being acted upon, since at that time among other things the capacity in the seeds is brought into play as the seeds are (as it were) warmed through. This is why all seeds do not sprout in response to the same cause, but only when the tempering of the air that causes this warming prevails.[1]

Now that this point has been laid down, it is evident that rapid sprouting is not a matter of strength but rather of weakness; for it is the weak thing that is readier to be acted upon[2] (which is why annuals are the more rapid sprouters; then too their nakedness is as it were a second factor contributing to their being acted upon).

The same holds for seed production: the weaker seeds are easier to mature[3] both for the air and for the starting-points in the plants themselves. This

1.4

---

[1] *Cf. CP* 1 10. 5–6.
[2] *Cf. CP* 1 10. 2.
[3] *Cf. CP* 2 11. 7.

---

[10] Wimmer : αὐταῖς U.
[11] Gaza (*enim vero*).
[12] Gaza (*eis*), Scaliger : αὐταῖς U.

THEOPHRASTUS

ζῴων ἐστίν· αἰεὶ γὰρ τὰ ἰσχυρότερα χρονιωτέρας[1]
ὡς ἐπίπαν ποιεῖται καὶ τὰς κυήσεις καὶ <τὰς>[1]
τροφάς. ἅμα δ' ὥσπερ ἐναντίως[2] τὰ μὲν αὐτοῦ
τοῦ χιτῶνος[3] [σπερματος][4] γεννητικά, σῶμά τε[5]
οὐδὲν <οὐ>[4] μικρὸν ὑπόκειται· τὰ δὲ ἐν ὄγκοις
μείζοσιν, οἳ καταναλίσκουσιν[6] εἰς ἑαυτοὺς[7] τὰς
τροφάς, ὥστε[8] τὰ μὲν ὀλιγόκαρπα, τὰ δὲ ὅλως
ἄκαρπα ποιεῖν, ὡς ἐπὶ τῶν ζῴων συμβαίνει τῶν
ἐκπαχυνομένων καὶ εὐτροφούντων.

1.5   τοῦ μὲν οὖν θᾶττον ἐκτελεοῦν[9] οὐχ ἡ ἰσχύς,
ἀλλὰ ταῦτα τὰ αἴτια, διὸ καὶ πολυγονώτερα.

καίτοι τάχ' ἄν τις ἀντείποι περὶ τοῦ πλήθους,
φάσκων πλείω τὰ ἀπὸ τῶν δένδρων εἶναι· πολλὰ
γὰρ ἕκαστον ἔχειν[10] τῶν περικαρπίων,[1] τὸ δὲ

---

[1] aP.
[2] ego : ἐὰν U.
[3] ego : χειμῶνος U.
[4] ego.
[5] ego : σωματα U.
[6] u : -ωσιν U.
[7] u : ἑαυτὰς U N aP.
[8] Gaza (*ut*), Itali : ὥσπερ U.
[9] U^c : ἐκτελοῦν
[10] U : ἔχει u. U^{ac}

---

[1] The pericarpia, hulls and stones.

DE CAUSIS PLANTARUM IV

holds of animals too: it is always the stronger ones that for the most part take a longer time not only for gestation but also for feeding the young. Then too the production is as it were of two opposite sorts: the annual seeds generate only a tunic, and nothing under the tunic is anything but small in body; whereas the seeds of trees are contained in adjuncts of larger bulk,[1] and these expend on themselves the food for the seeds, so that this makes some trees bear but little fruit[2] and other none at all,[3] as happens with animals that get fat and feed well.[4]

To conclude: the more rapid maturing of seed is 1.5 not due to strength, but to these reasons, which is why annuals are more prolific.

Yet one might deny this point about their prolific character and say that trees produce more seeds, since each pericarpion has many seeds,[5] and the

---

[2] That is, seed.

[3] *Cf. CP* 1 15. 3.

[4] *Cf.* Aristotle, *On the Generation of Animals,* i. 18 (725 b 29–34) [of animals and plants]: "... for some have much, some little, and some no seed at all not through weakness, but in some through the opposite; for it is expended on the body, as with some men: for being in a robust state and getting fleshy or too fat they expend seed less and desire intercourse less." *Cf.* also *ibid.* ii. 7 (746 b 24–29).

[5] This holds of vine, pear and apple, but not of olive, plum, date or the nuts (*cf. CP* 4 1. 2).

THEOPHRASTUS

δένδρον ἀφ' ἑνὸς γεγενῆσθαι σπέρματος.

ἀλλὰ ταῦτα μὲν τὰ ἐλαφρότερα, καὶ οἷον ἀπηρτημένα ἂν φανείη · τὸ δὲ μὴ δύνασθαι τηρεῖν τὰ[1] γένη, μηδ' ἐξομοιοῦν, ἐν[2] ἀμφοῖν ἂν ἔχοι τὸ αἴτιον, καὶ τῷ πλείω χρόνῳ[3] ὄντα[4] κατὰ γῆς μᾶλλον κατακρατεῖσθαι, καὶ τῷ τοὺς σωματικοὺς

1.6 ὄγκους ἀποσπᾶν.[5] ὃ γὰρ καὶ τῆς ἀκαρπίας αἴτιον, καὶ τῆς κακοκαρπίας εὐλόγως, φανερὸν δὲ μάλιστα ἐκ τῶν ἀμυγδαλῶν, εἴπερ ἀφαιρουμένης τῆς ὑγρότητος καὶ τῆς εὐτροφίας μεταβάλλουσιν.

τὰ δ' ἐπέτεια βραχύν τινα χρόνον ἐν τῇ γῇ γίνεται καὶ ὀλίγην ἕλκει τροφήν, διὸ καὶ οὐκ ἐξίσταται μιᾷ σπορᾷ τῶν γενῶν ἀλλὰ τῇ τρίτῃ,[6] τότε γὰρ ποιεῖται τὰς μεταβολάς · ὥσθ' ὅπερ ἐκείνοις διὰ τὸ πλῆθος τῆς τροφῆς εὐθύς, τοῦτο τοῖς σπέρμασιν χρονισθεῖσι κατὰ λόγον · πλὴν ὅτι τὰ μὲν

---

[1] U : Scaliger deletes : τάχ' Schneider.
[2] Wimmer (de Gaza : ἀπ' Schneider) : ἐπ U.
[3] u N P : χρόνωι U (-ῳ a).
[4] ego : τὰ U.
[5] U : ἀντισπᾶν (attrahat Gaza) Schneider.
[6] Gaza (sed tercio [sc. satu]), Dalecampius (ἀλλ' εἰ τῇ τρίτῃ Wimmer) : ἀλλ' ὁ τηι τρίτηι U: ἀλλ' ὅτι τρίτηι u : ἀλλ' ὁ τὴν τρίτην N : ἀλλὰ τὴν τρίτην aP.

---

[1] Unlike some vegetables, where several seeds are sown

tree was produced from a single seed.[1]

But these are the slighter parts of the difficulty, and would appear as it were peripheral. The main part of it is this: that the seeds of trees are unable to maintain their kind and unable to produce fruit like the parent; and the reason would lie in two circumstances: that they remain longer in the ground and so are mastered to a greater degree,[2] and that the great bulk of the pericarpion diverts the food.[3] For the reason for bearing no fruit at all is also, as we should expect, the reason for bearing inferior fruit.[4] That this is so is clearest in the case of almond-trees, where mutation follows as the fluidity and good feeding are eliminated.[5]

1.6

Seeds of annuals on the other hand remain but a short time in the ground and draw but little food, and for this reason do not depart from their kind in a single sowing, but in the third,[6] for it is then that they change over. And so it is reasonable that what happens to tree seeds directly, because of their great intake of food, should happen to annual seeds after an interval. But there is this difference: tree

---

together and said to produce a single plant (*cf. CP* 5 6. 9).

[2] This applies to the seed as producer of the tree.

[3] This applies to the tree as producer of the seed.

[4] The seed in the ground is overpowered by the fluid there. *Cf. CP* 1 17. 9–10.

[5] *Cf. CP* 1 17. 9–10.

[6] *Cf. HP* 8 8. 1, *CP* 1 9. 3, 2 13. 3.

THEOPHRASTUS

εἰς τὸ χεῖρον αἰεὶ μεταβάλλει, τὰ δὲ καὶ ἐπὶ τὸ βέλτιον, ἐὰν ἡ χώρα τοιάδε καὶ ὁ ἀήρ. ἀλλὰ δὴ τοῦτο μὲν ἕτερον.

1.7 εἰς ἄλλο δ' ὅλως ἐξίσταται γένος μάλιστα τῶν σπερμάτων τὰ ἰσχυρότερα, καθάπερ ὁ πυρὸς καὶ ἡ κριθή· μόνα γὰρ ἐξαιροῦνται, μᾶλλον δ' ὁ[1] πυρός, ἰσχυρότερον ὤν·[2] ὥστε καὶ τοῦτο ὁμολογούμενον τῷ[3] πλείω τροφὴν ἕλκειν. ὁ γὰρ θερμὸς ὥσπερ ἄπεπτόν τι τὸ ὅλον.

2.1 τὰ μὲν οὖν κατὰ τὴν βλάστησιν ἐν τούτοις ἂν εἴη.

πρὸς δὲ τὴν ἔξω διαμονὴν φυτευομένων,[4] ἄλλοις μὲν ἄλλως συμβαίνει κατὰ τοὺς χρόνους. οὐ μὴν ἀλλ' οὐδ' ἄλογον ἔνια προτερεῖν τῶν δενδρικῶν εἰς τὴν φθοράν, οἷον ὅσα ἐν περικαρπίοις ἐστὶ

---

[1] u : διὸ U.
[2] u : ὄν U.
[3] τῶι U (τῷ N P) : τὸ u a.
[4] U : Gaza omits : τῶν σπερμάτων Schneider : φυλαττομένων Wimmer.

---

[1] It was held that imported seed changed to the native variety. From this Theophrastus infers that if the native variety was better the change was an improvement, and due to the change of country and climate.

seeds always change for the worse, whereas the seeds of annuals also change for the better, if the country and weather have the right character.[1] But improvement is another matter.

Among seed crops the change to a quite different plant occurs chiefly in the stronger ones, as wheat and barley, for they alone turn into darnel,[2] and wheat, the stronger of the two, does so more. So here too is a case that agrees with the explanation of degeneration as due to taking in too much food. As for lupine,[3] it is (so to speak) a thing not concocted at all.

And so matters concerned with the sprouting of the seeds would have their explanation in the points mentioned.

Survival of the Seed: Trees

The survival of the seed outside the ground for purposes of propagation varies in length of time for the seeds of different plants. Nevertheless it is not unreasonable that some tree seeds should perish earlier than annual seeds, such as all that are contained in fleshy pericarpia when these (for instance)

1.7

2.1

---

[2] *Cf. HP* 2 4. 2; 8 5. 1; 8 8. 3.

[3] Lupine, which like wheat and barley is very strong (*cf. HP* 8 11. 8), nevertheless never changes over to another plant. Its powers of concoction are not overwhelmed by too much food, since it has no such powers to speak of.

THEOPHRASTUS

σαρκώδεσιν,[1] οἷον[2] ὅταν χωρισθῇ ἢ[3] σαπῇ, γυμνούμενα γὰρ ἀναξηραίνεται καὶ θνήσκει· τὰ δὲ ἐν δερματικοῖς καὶ μὴ χωριζόμενα, καὶ γὰρ ὑγρότερα καὶ παρεισδέχεται τὸν ἀέρα. μάλιστα δὲ διαμένει τὰ ἐν ξυλώδεσιν, καὶ τούτων ὅσα πυκνόν τε[4] τὸ κέλυφος ἔχει, καὶ αὐτὰ λιπαρά, καθάπερ τὰ κάρυα τὰ Ἡρακλεωτικά, πλὴν ἐκπικροῦται·[5] τὰ δ' ἀμύγδαλα ἐλάττω χρόνον, ἐλάχιστον[6] δὲ τὰ βασιλικά, [μόνον][7] μανότατον γὰρ τὸ πέριξ καὶ ἥκιστα συμφυές. διαμένει δὲ καὶ τὸ τῆς ἐλάας καὶ εἴ τι ἄλλο τοιοῦτον. τῶν δ' ἐν σαρκώδεσι περικαρπίοις χρονιώτατον τὸ τοῦ φοίνικος, ἅτε ξηρότατον ὂν καὶ πυκνόν.

2.2 τὰ μὲν οὖν τῶν δένδρων διὰ ταύτας τὰς αἰτίας τὰ μὲν μᾶλλον, τὰ δ' ἧττον διαμένει.

σχεδὸν δὲ καὶ τὰ[8] τῶν σιτωδῶν διὰ παραπλησίας αἰτίας· ἢ γὰρ τῷ περιέχεσθαι πλείοσι χιτῶσιν ἡ διαμονή, καθάπερ ὁ κέγχρος, ἢ τῷ λίπος

---

[1] u : σαρκώδεις U. [2] [οἷον] Wimmer.
[3] u (ἢ U) : καὶ Wimmer.
[4] Schneider : τι U.
[5] u : ἐκ πικροῦ U.
[6] U^c : ἐλαχίστουν U^{ac}.
[7] ego : μένει Wimmer. [8] U^c : ὄντα U^{ac}.

become detached from the seed or decompose, since the seed on being exposed dries out and dies. The seeds in leathery pericarpia perish even when not detached, since the pericarpia are less rigid than woody ones and admit the air. Seeds in woody pericarpia survive best, and of these all that have a close-textured husk and are themselves oily, as filberts (except that they turn bitter). But almonds last a shorter time, and walnuts the shortest time of all, since the covering is of very open texture and least of all coalesces into a single shell. Olive seeds and the like[1] also survive, but of seeds in a fleshy pericarpion the longest lasting is the date seed, since it is extremely dry and of close texture.[2]

In trees, then, some seeds last longer, some less, for these reasons.

2.2

Survival of the Seed: Seed Crops

So too (one might say) the survival of the seeds of seed crops is due to similar reasons: either the seeds are wrapped in a number of coats, as millet,[3] or are

---

[1] Where the germ is in a woody "stone," and this is in a fleshy pericarpion: *cf. CP* 4 1. 2.

[2] *Cf. CP* 4 1. 2 with note 1.

[3] *Cf. HP* 8 11. 1: "The seeds of seed crops do not have the same power of sprouting and of keeping. For some sprout and mature very rapidly and keep very well, as Italian millet and millet . . ."

ἔχειν, ὥσπερ τὸ σήσαμον, ἢ τῷ δριμύτητά[1] τινα καὶ πικρότητα χυλοῦ, καθάπερ ὁ θέρμος καὶ ὁ ἐρέβινθος καὶ ὁ ὄροβος· μόνα γὰρ δὴ καὶ οὐ ζῳοῦται τῶν χεδροπῶν ὁ θέρμος καὶ ὁ ἐρέβινθος,[2] ἀλλ᾿ ὅ γε ἐρέβινθος μέλας γίνεται διαφθειρόμενος. ὁ δὲ πυρὸς μᾶλλον τῆς κριθῆς καὶ τῶν χεδροπῶν, διὰ τὸ θερμότερον εἶναι καὶ χιτῶνας ἔχειν πλείους· ἡ γὰρ αὖ κριθή, καθάπερ γυμνόν· τὰ δὲ χεδροπὰ παχέα μὲν τοῖς κελύφεσιν, ἀλλὰ μανά, καὶ γλυκύτητά τινα ἔχοντα φαίνεται· οὐχ ἧττον αἴτια ταῦτα τῆς φθορᾶς, διὸ καὶ ὁ κύαμος καὶ ὁ ὦχρος τάχιστα κόπτεται. τάχα δ᾿ ἀληθέστερον εἰπεῖν ὡς κατὰ τὰς χώρας· ἐν Ἀπολλωνίᾳ γοῦν τῇ περὶ τὸν Ἰόνιον πολλά φασιν ἔτη διαμένειν τοὺς κυάμους, πολλὰ δὲ καὶ περὶ Κύζικον.

ἀλλὰ περὶ μὲν τῆς τούτων διαμονῆς, καὶ τὸ ὅλον τῆς φύσεως, τάχ᾿ ἂν ἐν τοῖς ὕστερον ἐπὶ

---

§ 2 lines 7–10 Athenaeus epitome ii. 45 (55 E): Θεόφραστος δὲ ἱστορεῖ ἐν αἰτίοις φυτικοῖς ὅτι θέρμος καὶ ὄροβος καὶ ἐρέβινθος μόνα οὐ ζῳοῦται τῶν χεδροπῶν διὰ τὴν δριμύτητα καὶ πικρότητα. ὁ δ᾿ ἐρέβινθος, φησί, μέλας γίνεται διαφθειρόμενος.

[1] U^c : δριμύτατον U^{ac}.
[2] u : ῥέβινθος U.

DE CAUSIS PLANTARUM IV

oily, as sesame, or have a certain pungency and bitterness of flavour, as lupine, chickpea and vetch,[1] for alone among legumes lupine and chickpea survive without breeding animals[2]; instead when chickpea perishes it turns black. Wheat lasts better than barley and legumes because it is hotter and has more coats; barley on the other hand is practically naked, and legumes, although thickly wrapped in their pods, are nevertheless of open texture and are observed to possess a certain sweetness, and open texture and sweetness are no less responsible for destruction than other causes, which is why bean and birds' pease get wormy soonest. Perhaps it is truer to say that survival goes with the country: thus at Apollonia on the Ionian Sea it is said that beans keep for many years, and also at Cyzicus.[3]

The survival of seed crops and their nature in general will perhaps be discussed later at greater

---

[1] *Cf. HP* 8 11. 2.

[2] *Cf. HP* 8 11. 2: "In seed crops as they perish peculiar animals are bred ... except for chickpea, for this alone breeds none." *Cf.* also *HP* 8 11. 6 for chickpea, lupine, vetch and millet as long-lasting seeds.

[3] *Cf. HP* 8 11. 3: "Country and air make a difference in whether seed crop seeds get worm-eaten or not. Thus at Apollonia on the Ionian Sea they assert that bean does not get worm-eaten at all, and is therefore stored for keeping; and it also keeps for a considerable time in Cyzicus."

THEOPHRASTUS

πλέον ῥηθείη · πρὸς δὲ τὰ τῶν δένδρων συγκρινόμενα, ταύτας[1] ἔχει τὰς διαφοράς.

3.1 ἐν δὲ τοῖς τῶν λαχανωδῶν[2] τὰ μὲν ἄλλα τὴν ἐξομοίωσιν ἀποδίδωσιν (ῥάφανος γὰρ[3] καὶ ἄλλ' ἄττα δοκεῖ παραλλάττειν, ἃ[4] καὶ δενδρικώτερα), τὰς δ' ἐκφύσεις πλέον ἀλλήλων ἔτι[5] ταῦτα παραλλάττει[6] τῶν σιτηρῶν · τὰ μὲν γὰρ τριταῖα διαβλαστάνει, καθάπερ ὤκιμον σίκυος κολοκύντη, τὰ δὲ πεμπταῖα ἢ ἑκταῖα, τὰ δὲ πεντεκαιδεκαταῖα, καθάπερ πράσον, τὸ δὲ σέλινον τεσσαρακοσταῖον, ἐνιαχοῦ δὲ πεντηκοσταῖον,[7] δυσφυέστατον γὰρ τοῦτο πάντων. δυσφυὲς δὲ καὶ τὸ κορίαννον,[8]

---

[1] u : τασταs U.
[2] U^cc : λαχανῶν U^ac.
[3] U : Gaza omits : δὲ Schneider.
[4] ego (παραλλάττειν Wimmer) : γὰρ ἀλλάττειν ἃ U.
[5] Itali : ὅτι U.
[6] Gaza, Scaliger : -ειν U.
[7] aP (σ smudged) : -κοτ- U N.
[8] τὸ κορίανον u : τορίανον U (τὸ ῥιανὸν N aP).

DE CAUSIS PLANTARUM IV

length[1]; meanwhile the seeds as compared with those of trees show these differences.

Vegetable Seeds: Sprouting

Among the seeds of vegetables the rest breed true (I say "the rest" since cabbage[2] and certain other vegetables are considered to vary, and these are precisely the vegetables with the greatest resemblance to trees)[3]; but in the time taken for coming up these differ among themselves even more than grains.[4] For some come up on the third day after sowing, as basil, cucumber and gourd; some on the fifth or sixth, some on the fifteenth, as leek; whereas celery comes up on the fortieth day and in some places on the fiftieth, since it is the slowest of all to grow. Coriander is also slow, for the seed does not sprout

3.1

---

[1] *CP* 4 15. 3–4 16. 2 (for survival); *CP* 4 7. 4–7 (for nature).

[2] It was often planted from side-growths: *cf. HP* 7 2. 1; *CP* 1 4. 2.

[3] *Cf. HP* 1 3. 4: "Thus of undershrubs and vegetables some turn out to have a single trunk and as it were the nature of a tree, as cabbage [so Gaza; ῥαφανίς U] and rue, and for this reason some persons call such vegetables 'tree-vegetables'; ..."

[4] *Cf. HP* 8 1. 5 (wheat and barley come up on the seventh day, pulses [except bean] on the fourth or fifth; bean on the fifteenth or even on the twentieth).

THEOPHRASTUS

οὐ γὰρ βλαστάνει μὴ ἐριχθέν ·[1] αἰτία δ' ἡ σκληρό-
3.2 της, ὥσπερ γὰρ ξυλῶδές ἐστι τὸ περιέχον. ἡ δὲ
τῶν τευτλίων ἀνωμαλία, τῷ[2] τὰ μὲν ὕστερον
μηνὶ[3] βλαστάνειν,[4] τὰ δὲ δυοῖν, τὰ δὲ πλείοσιν,
τὰ δὲ καὶ ἐνιαυτῷ, σημαίνειν ἔοικεν τῶν σπερμάτων αὐτῶν ἀτεραμνότητά τινα πρὸς τὴν
βλάστην.[5]

τὰ δὲ θερινὰ σπειρόμενα[6] δῆλον ὡς δι' ἀσθένειαν · διὸ ταχύ τε παραγίνεται καὶ ταύτην τὴν
ὥραν φύεται, φέρειν οὐ δυνάμενα τὸν χειμῶνα,

---

[1] ego (cf. HP 7 1. 3 ἐλιχθῇ U) : ἐρεχθὲν U.
[2] τῶι U^r (τῶ N aP) : τῶν U^{ar}.
[3] Gaza (*mense*), Schneider : μὴ U N^c (N^{ac} omits) : μηνὸς u aP.
[4] ego (ἐκβλαστάνειν Wimmer) : ἐπιβλαστάνειν U.
[5] U N : βλάστησιν aP.
[6] u : πειρόμενα U.

---

[1] *Cf. HP* 7 1. 3: "Not all herbaceous plants come out of the seed in the same time ... The quickest to do so are basil, blite and rocket, and, among those sown for winter, radish, for they come up more or less on the third day. Lettuce comes up on the fourth or fifth; cucumber and gourd in about five or six days, some say seven, cucumber coming up earlier and faster; purslane in more; dill on the fourth day, cress and mustard on the fifth, beet in summer on the sixth, in winter on the tenth; orach on the

unless it is bruised.[1] The reason is its hardness, for the casing is like wood. On the other hand the irregularity in beets, some coming up a month later than the rest, some two, some several months, and some even a year,[2] appears to indicate a certain intractability to sprouting on the part of the seeds themselves.

3.2

Seeds sown in the summer crop[3] are evidently sown at this time because of weakness, which is why they come up rapidly and grow in that season, since they are unable to endure winter, some of them

---

eighth, cabbage on the tenth. Leek and horn-onion differ in the time taken, leek coming up on the nineteenth day and in some places on the twentieth, horn-onion on the tenth or twelfth. Coriander is slow, for fresh seed will not sprout unless it is bruised [reading ἐριχθῇ for ἐλιχθῇ U]. Savory and marjoram take longer than thirty days, and celery is slowest of all: those who give the shorter period say it comes up on the fortieth, whereas others say the fiftieth..."

[2] *Cf. HP* 7 1. 6: "They say that there is a peculiarity in beet: not all the seed comes up at once, but some comes up much later, and some in the next year and the third, which is why little comes up from much seed."

[3] *Cf. HP* 7 1. 1–2 (the three seasons of sowing vegetables are [1] that of the winter crop or first crop, sown after the summer solstice in Metagitnion [August], [2] that of the second crop, sown after the winter solstice in Gamelion [January], and [3] that of the summer crop, sown in Munychion [April]): "in this crop is sown cucumber, gourd, blite, basil and purslane."

τὰ μὲν[1] ὄντα ξηρά, καθάπερ καὶ τὸ ὤκιμον, τὰ δὲ ὑγρὰ καὶ ψυχρά, καθάπερ ὁ σίκυος καὶ ἡ ἀνδράχνη.[2]

θαυμαστὸν δ' ἂν δόξειεν τὸ τοῦ ἀβροτόνου[3] μάλιστα διότι θερμὸν ὂν τὴν ἀλέαν διώκει· αἰτία δὲ ἡ ἀσθένεια· πρὸς ἄμφω γὰρ ἀσθενεῖ,[4] καὶ πρὸς τοὺς χειμῶνας καὶ πρὸς τὰ καύματα.

3.3 τὸ δ' ὅλον οὐκ ἔοικεν[5] ἡ θερμότης μάλιστα εὐβλαστεῖν, εἴπερ θερμὰ τὰ δριμέα· καὶ γὰρ τὸ πράσον καὶ τὸ γήτειον καὶ ἔτι μᾶλλον ἡ θύμβρα καὶ ἡ ὀρίγανος δυσβλαστῆ· δεῖ γὰρ ἔχειν τινὰ ὑγρότητα, καὶ οὐχ ἧττον, ἀλλὰ μᾶλλον ἴσως, εὐκρασίαν τὸ εὐβλαστὲς πρὸς τὸ ποιεῖν καὶ πάσχειν. ὅλως δ' (ὡς γένει[6] λαβεῖν) ξηρότατα τῶν σπερμάτων τὰ στεφανωματικὰ[7] καὶ τὰ τῶν λαχάνων, ὅθεν καὶ τάχιστα τὰς ἰκμάδας ἕλκει, διὸ καὶ κρεμαννύουσιν αὐτὰ καὶ οὐ ῥαίνουσιν τὰ οἰκήματα, οὐδ' ὕδωρ εἰσφέρουσιν ἁπλῶς.[8]

---

[1] Itali (Gaza omits) : μὴ U.
[2] aP : ἀνδραχύνη U N.
[3] U^r N aP : ἀμροτόμου U^{ar}.
[4] u : ἀσθενῆ U.
[5] U : οὐ ποιεῖ Wimmer (but cf. εὐβλαστής of ἀήρ CP 2 3. 3).
[6] <ἐν> γένει u.
[7] Basle ed. of 1541 : στεφανωτικὰ U.
[8] U : ὅλως Wimmer.

DE CAUSIS PLANTARUM IV

being dry (as basil among others), some fluid and cold (as cucumber and purslane).

The case of tree-wormwood[1] would appear most surprising, because the plant is hot and yet seeks warmth. The reason is its weakness, for the plant is too weak not only for cold weather but for hot.

In fact it does not appear that heat is the greatest factor in quick emergence, if we take pungent plants to be hot, since leek, horn-onion, and still more savory and marjoram are slow. For to come up rapidly a seed must possess a certain fluidity, and no less than this, or rather perhaps even more, it must have a certain good tempering that fits it to act and to be acted upon. In general the driest seeds as a class are those of coronary plants and vegetables, which is why they are the quickest to attract moisture, and for this reason they are hung up away from the ground and the rooms are not sprinkled or any water brought into them at all.[2]

3.3

---

[1] *Cf. HP* 6 7. 3: "Tree-wormwood sprouts better from seed than from a root or a side-growth, but it grows with difficulty even from seed. They root the slip in pots first, like the gardens of Adonis, in summer, for it is very sensitive to cold and in a word delicate even in a spot where the sun reaches it very bright, but when it has taken and grown it is tall and strong and tree-like ..." The comparison with the gardens of Adonis indicates that the pots were indoors; hence the plant can be said not to endure hot weather.

[2] *Cf. CP* 1 7. 2.

THEOPHRASTUS

3.4 ἡ δὲ διαμονὴ θησαυριζομένων[1] πρός τε τοὺς σπόρους[2] καὶ πρὸς τὰς ἄλλας χρείας παραπλησία[3] καὶ τοῖς σιτηροῖς· γόνιμα μὲν εἰς τετραετίαν[4] μάλιστα, χρήσιμα δὲ πρὸς τἆλλα πλείω χρόνον, ὥσπερ ὁ σῖτος εἰς τροφήν. καὶ τοῦτο δὲ καὶ[5] κατὰ λόγον, ὥσπερ καὶ τοῖς ᾠοῖς[6] πρῶτον ἀπολιπεῖν[7] τὴν τῆς γονῆς.[8] καὶ γὰρ ἐν τούτοις ἡ ἀρχὴ τοῦ γεννᾶν, ὅπερ ἄν τις ὡς σπέρμα θείη, τὸ δὲ λοιπὸν ὡς τροφὴν καὶ ὕλην προσηρτημένην,[9] ἀλλὰ τὸ ὅλον καλεῖται σπέρμα· διὸ καὶ ζῆν αὐτά φασι καὶ μὴ ξῆν ὅταν διαμένῃ ταύτῃ ἢ φθαρῇ[10] (καθάπερ καὶ τὰ ᾠά[11])· συμβαίνει δὲ καὶ τοῦτο

---

[1] Wimmer : -νη U.
[2] u : πόρους U.
[3] u : -ήσια U N aP.
[4] U^cc (from τεσ-).
[5] U N : aP omit.
[6] ego : ζῶοις U.
[7] U : -λείπειν Schneider.
[8] γονῆς <δύναμιν> Schneider (vis generandi Gaza).
[9] N aP : -στηρ- U.
[10] Schneider : φθαρείη U.
[11] Gaza (ova) : ἴα U.

---

[1] *Cf. HP* 8 11. 5: "For coming up and for sowing in general the best (*sc.* seeds in cereals) are considered to be

DE CAUSIS PLANTARUM IV

Vegetable Seeds: Survival

The survival of the seeds when stored for sowing 3.4
and for other uses is similar to that of cereals,[1] the
seeds remaining fertile for about three years[2] and
good for other purposes still longer, as cereals are
good for food. This is reasonable in the seeds too,
just as it is reasonable in eggs that their use for propagation should leave them first.[3] For in seeds too
there is the starting-point of generation, which one
would count as the true "seed," accounting the rest
as food and matter attached to this.[4] The whole
complex, however, is termed "seed," which is why
people speak of the seeds as "living" or "not living"
when they survive or perish in this respect (just as
they say this of eggs)[5]; and this occurs not only in

---

those of last year. Those of two and of three years ago are
inferior, and those older than this are (one may say) infertile, though good enough for food."

[2] *Cf. HP* 7 5. 5 (of vegetable seeds): "None keeps longer
than four years so as still to be good for sowing."

[3] *Cf.* Aristotle, *On the Generation of Animals*, ii. 5 (741 b
19–22): ". . . it happens in all cases that what is produced
last fails first, and what is produced first fails last, as if
nature were taking the home lap and being resolved back
to the starting-point from which it began."

[4] *Cf. CP* 1 7. 1.

[5] *Cf.* Aristotle, *On the Generation of Animals*, ii. 5 (741 a
19–20) [fertile eggs can be said to live because an actual-
ized animal being could have come from them].

THEOPHRASTUS

καὶ τοῖς βαλανώδεσιν καὶ τοῖς ἄλλοις.

εἰς βλάστησιν δὲ τὰ ἕνα τῶν νέων οὐκ ἄλογον εἶναι βελτίω, ξυνεστηκότα τε καὶ οἷον αὐτὰ αὑτῶν ὄντα μᾶλλον · πρόσεστι γάρ τις καὶ τούτων[1] πέψις,[2] ἀποπνεύσαντος τοῦ ἀλλοτρίου, χρονιζομένων[3] δὲ πάλιν γῆρας καὶ φθίσις.[4]

3.5 περὶ δὲ τοῦ ἐκκαυλεῖν τάχιστα μὲν τὰ ἀπὸ τῶν ἀκμαζόντων, ὡς ἰσχυροτάτων (τελέωσις γάρ τις ἡ ἐκκαύλησις, εἴπερ ὁ καρπὸς τέλος) οὐκ ἄλογον · δευτέρα[5] δὲ ἡ ἐκ τῶν παλαιοτέρων, καὶ γὰρ ἐκ

---

[1] U : τούτοις Schneider.
[2] U r N aP : πέμψις U ar.
[3] Schneider : -νου U.
[4] U c : -ιν U ac.
[5] u P : δευτερα U : δεύτερα N a.

---

[1] Acorn-like seeds resemble certain eggs in having a shell and in being rounded at one end and tapering at the other.
[2] For the phrase cf. CP 3 4. 4; 3 7. 10.
[3] Cf. CP 2 8. 3.
[4] Cf. HP 7 3. 1, 7 3. 4 (of vegetables): "Those from older seeds send up a stem sooner, but soonest to do so are plants from seeds in their prime, for seeds too have their prime."

DE CAUSIS PLANTARUM IV

acorn-like seeds but also in the rest.[1]

Vegetable Seeds: Rapidity of Germination

It is not unreasonable that, for purposes of sprouting, last year's seeds should be better than this year's, since they are firmer and (as it were) more their own selves.[2] For even in the kept seed there is a certain concoction when the foreign ingredient has evaporated off[3]; but when the seed is kept still longer we have instead old age and death.

As for plants coming from seeds in their prime being the quickest to send up a stem,[4] there is nothing unreasonable here, these seeds being at their strongest (since to develop a stem is a kind of achievement of the goal, if the fruit[5] is taken as the goal).[6] Second in rapidity is the production of a stem from older seeds,[7] since the power proceeding

3.5

---

[5] "Fruit" (*karpós*) is often used of a fruiting stem or stalk.

[6] In most vegetables the leaves, and not the seed or fruit, are the end so far as human consumption is concerned.

[7] *Cf. HP* 7 1. 6 (of herbaceous plants): "A difference in rapidity of sprouting is also made by the age of the seeds. Some come up more rapidly from young seed, as leek, horn-onion, cucumber and gourd ...; others from old seed, as celery, beet, cress, savory, coriander and marjoram (that is, if again [reading αὖ for οὐ of U] the plant does not come up more rapidly from seed that is young, as we said)."

215

THEOPHRASTUS

τούτων οἷον καθαρωτέρα τις καὶ πλείων ἡ δύναμις, ἐν δὲ τοῖς νέοις ἀναμεμιγμένη,[1] καὶ τὸ πλέον εἰς τὴν τροφὴν ἄγουσα, καθάπερ ἐν τοῖς ἄλλοις· αἰεὶ γὰρ ὀψιαίτερα τὰ πολύτροφα· διὸ καὶ εἰ μὴ τὰ ἀπὸ τῶν ἀκμαζόντων ἐκκαυλεῖ πρῶτα, ἀλλὰ τὰ ἀπὸ τῶν παλαιοτέρων, οὐκ ἄλογον.

3.6 περὶ οὗ δὴ καὶ ἀντιλέγουσί τινες· ἐκκαυλεῖν γάρ φασι[2] τὰ ἀσθενέστερα μᾶλλον, ὥσπερ ἐπὶ τῶν λαχάνων, αἴτιον δέ, ὅτι εὐπαθέστερά τε, καὶ ἐλάττω τροφὴν ἔχοντα τελειοῖ (τὸ δ' ἐκκαυλεῖν ὥσπερ τελέωσιν[3]). διὸ κἀκ τῶν δένδρων τὰ πρεσβύτερα θᾶττον πέττει καὶ μᾶλλον τοὺς καρπούς, τὰ δὲ ἐν ἀκμῇ πλείους μὲν <καὶ>[4] καλλίους ἔχει, βραδύτερον δέ.

φαίνεται δ' οὖν, εἰ τοῦτο ἀληθές, ὅτι ταχυγονώτερα τὰ παλαιότερα· μεμῖχθαι γάρ τις ἐν τοῖς

---

[1] u : -ην U.
[2] U^c : φησι U^{ac}.
[3] U : τελέωσις Heinsius.
[4] a.

---

[1] For animals cf. Aristotle, On the Generation of Animals, i. 18 (725 b 19–25): "Further, semen is found neither in the earliest age nor in old age nor in illnesses, ... in youth because of growth, for everything is first used up on that, since in about five years in the case of man at least the body is held to acquire half of the whole size

DE CAUSIS PLANTARUM IV

from these too is purer (as it were) and greater, whereas in young seeds it is mixed and more conducive to feeding than to creation, just as it is in other young parts (since whatever does much feeding is always later in development).[1] This is why it is not unreasonable that if the plants from seeds in their prime are not first to send up a stem, it should be the plants from older seed that do so instead.[2]

Some dispute this and assert that it is rather the weaker seeds that are quickest to send up stems (as in vegetables), the reason being that the weaker seeds are more easily affected and having less food bring it to completion (sending up a stem being, they say, an achievement as it were of the goal); this moreover is why in trees the older ones concoct their fruit quicker and better, whereas trees that are in their prime have more and finer fruit[3] but take longer to get it.

3.6

In any case it appears (if this is true) that it is the older seeds that generate more rapidly. For it is considered[4] that matter of a sort and food for the

---

achieved during the rest of the time." (For this last point *cf.* Plato, *Laws* vii 788 D 5–8.)

[2] *Cf.* note 7, p. 215. [3] *Cf. CP* 2 11. 10.

[4] *Cf. CP* 4 3. 4; 1 7. 1 and Aristotle, *On the Generation of Animals*, i. 23 (731 a 5–9): "For the egg is a fetation and the animal comes from part of it, the rest being food, and the plant too comes from part of the seed, the rest becoming food for the sprout and the first root."

THEOPHRASTUS

σπέρμασιν οἷον ὕλη δοκεῖ[1] καὶ τροφὴ ταῖς ἀρχαῖς ἣν δεῖ[2] προκατειργάσθαι πρότερον ἢ εἰς τὴν βλάστησιν ἐλθεῖν.

3.7 εἴη δ' ἂν κἀκεῖνο λέγειν πρός γε τὸ ἐκκαυλεῖν ὡς, ἧττον ῥιζουμένων τῶν παλαιῶν, εἰς δὲ τὸ ἄνω μᾶλλον[3] φερομένων, ταχεῖαν ποιεῖται[4] τὴν τελείωσιν, ὥσπερ σχεδὸν καὶ ἐπὶ τῶν ὀλιγοχρονίων πάντων ἐστίν (καθάπερ καὶ ἐπὶ τῶν τριμήνων λέγεται). τοὺς δὲ χρόνους δῆλον ὅτι καὶ τὰς ἀκμὰς κατὰ τὰς φύσεις ἑκάστων ληπτέον.

ἀλλὰ γὰρ ταῦτα μὲν ὁμοιότητά τινα ἔχει· τὰ δ' ἴδια καθ' ἕκαστον γένος αὐτὰ καθ' ἑαυτὰ λέγουσιν ἐπὶ πλέον· ῥητέον δὲ τοῖς <τὰ>[5] τῶν δένδρων καὶ πρὸς αὐτὰ[6] καὶ πρὸς τὰ ἄλλα θεωροῦσιν.

4.1 ἄτοπον δ' ἂν δόξειεν καὶ ἅμα θαυμαστόν, εἰ ἀπὸ τῶν ἀτελῶν ἔνια γεννᾷ, καὶ ταῦτα δένδρων φύσεις[7] μεγάλας οὕτως οἷον ἰτέας καὶ πτελέας·

---

[1] οἷον ὕλη δοκεῖ ego : δοκεῖ · οἷον ὕλη
[2] U^c : δὲ U^{ac}.
[3] Gaza (*sursum potius*), Itali : ἀνώμαλον U.
[4] ego : ποιεῖ U.
[5] Heinsius.
[6] Schneider : αὐτὰ U.
[7] U^c : φύσειν U^{ac}.

starting-point is mixed in with the seed, and this food must first be worked up before the seed can reach the stage of sprouting.

One could make a further point, at any rate in support of the view that it is the plants from older seeds that first send up a stem,[1] that since the older seeds run less to root and more to upward growth, the plants reach their completion sooner, just as this direction is taken by practically all plants of short duration (as is also said of three-months cereals). We must evidently take length of time of sending up a stem and the prime of the seed in each case with reference to the nature of the particular plant in question.

3.7

But enough. These matters involve a certain similarity between different kinds of plant, whereas the authorities dwell more on special features, taken by themselves, of the various kinds. But the points must be discussed by those who study the seeds of trees as compared among themselves and with the seeds of other plants.

Tree Seeds: Peculiar Features

It would appear odd and wonderful as well that some trees should propagate from immature seed, and propagate moreover trees that grow as large as

4.1

---

[1] The discussion is not concerned with the objector's parallel with older trees (*CP* 4 3. 6).

THEOPHRASTUS

ἀτελὲς γὰρ τὸ ὠμόν. ἆρ' οὖν, εἴπερ τοῦτ' ἀληθές, διαιρετέον τὴν τελειότητα τήν τε πρὸς ἡμᾶς καὶ πρὸς τὴν γένεσιν; ἡ μὲν γὰρ πρὸς τροφήν, ἡ δὲ πρὸς δύναμιν τοῦ γεννᾶν· ἔνια δὲ ἄτροφα, γεννητικὰ δέ, τὰ δ' ἴσως ἀνάπαλιν. τάχα δ' ἐκείνη[1] πρός γε τὰ νῦν ἡ διαίρεσις, ὅτι τὴν πέψιν τιθέμεθα χρώμασιν καὶ χυλοῖς καὶ πυκνότητι[2] καὶ τοῖς τοιούτοις· ἐπεὶ τό γ' ἐδώδιμον ὑπάρχει καὶ τοῖς τῆς πτελέας καὶ ἄλλοις· ἀλλ' οὐ τοῦτο κύριον, ἀλλὰ τὸ γεννᾶν· ἕκαστον γὰρ τῷ ἔργῳ κρίνεται. καὶ ταῦτα μὲν ὡς πρὸς ὑπόθεσιν.

4.2 ἐκεῖνο[3] δὲ ἀτοπώτατον, εἰ τελεούμενα γένους τινὸς ἄγονα γίνεται τὸ ὅλον, καίτοι[4] ἀπὸ σπέρματος γινομένου τοῦ δένδρου, καθάπερ ἐπὶ τῆς κυπαρίττου· τὸ γὰρ ἄρρεν γένος ὅλως ἄγονον,

---

[1] ego : δ' ἐκείνη U^c : δὲκεινη U^ac.
[2] U^r aP : -τα U^ar N.
[3] U^r N aP : ἐκείνω U^ar.
[4] u : καὶ | τωι U.

---

[1] *Cf. HP* 3 1. 2–3: "All trees that have seed and fruit ... can also propagate from these; indeed authorities say that trees reputed to have no fruit propagate in this way, as

willows and elms, since a raw fruit is immature.[1] If this is true shall we distinguish two kinds of maturity, maturity for man and maturity for propagation? For the former kind provides food, the latter the power of generation, and some seeds are useless as food, but capable of generation, whereas others perhaps are the reverse.[2] But perhaps the distinction is to be made in the present circumstances as follows: that to us concoction lies in the colour and flavour and firmness and the like, since simple edibility is found not only in the seeds of the elm[3] but also in others. But what determines maturity is not such concoction, but propagation, since each part is judged by its function.[4] These remarks are made on the assumption that the raw seed generates.

But another matter is oddest of all: if trees of a given variety as they mature turn out to be entirely infertile, although the tree is produced from seed, as with cypress, for the whole male variety is infertile

4.2

---

willow and elm ... The truth, they say, is that the willow drops its fruit before making it vigorous and concocting it ... And that this is true of the elm they show as follows ..."; *HP* 3 3. 4: "The rest all bear fruit, but as we said, this is disputed in the willow, black poplar and elm."

[2] Useful as food but incapable of generation.

[3] Which are edible when raw.

[4] *Cf.* Aristotle, *On the Parts of Animals*, i. 1 (640 b 35–641 a 5).

THEOPHRASTUS

πολλὰ[1] δὲ καὶ τῶν θηλειῶν,[2] ἐπεὶ τό γε μὴ
καρποφορεῖν ἔνια τῶν ὁμογενῶν ἧττον ἄτοπον.

καὶ πάλιν ἐπὶ τῶν σπερμάτων αὖ, τὸ[3] πηρωθὲν
ἄγονον, ἀλλ' ἴσως τοῦτό γε καὶ ἀναγκαῖον · τὸ δὲ
μηδὲν ὅλως γόνιμον ὥσπερ ἐλέγχει τὴν φύσιν ὅτι
καὶ μάτην, ὃ καὶ ἡμῖν ὑπεναντίον πρὸς τὰ πρότε-
4.3 ρον. καὶ τὰ μὲν τῆς πτελέας ἀμφισβητούμενα καὶ
τῆς ἰτέας · ὁ δ' ἐρινεὸς ὁμολογουμένως γεννᾷ καὶ

---

[1] πολλα U : πολλαὶ Schneider.
[2] aP : θηλείων U N.
[3] αὖ τὸ u : αὐτὸ U N aP.

---

[1] For male and female cypresses cf. HP 1 8. 2; 2 2. 6; cf. also HP 1 14. 5: "Here is another peculiar feature separating the wild trees from the cultivated: people distinguish the wild into only or mainly the male and female variety, whereas they distinguish the cultivated into several varieties"; HP 3 3. 6–7: "At all events it is held that also with other plants of the same kind and called by the same name there is a division into a variety that has no fruit and another that has ... And just about all the trees called male among those of the same kind bear no fruit ..."

[2] Cf. CP 2 6. 1; 2 11. 2; 2 11. 7; 2 13. 5.

[3] CP 1 1. 1.

[4] Or "fruit"; they are the same. Aristotle speaks of the wild fig as having no fruit: On the Generation of Animals, i. 1 (715 b 21–25): (Testacea and other animals, because their nature is similar to that of plants, have no distinction of female and male any more than plants have, and when

DE CAUSIS PLANTARUM IV

(and so too are many of the females)[1]; as for the failure of occasional individuals of the same kind to bear fruit, there is less oddity here.

Again (to return to seeds) the mutilated individual is infertile, but perhaps the infertility here is a matter of mechanical causation[2]; whereas that no seed at all should be fertile as it were convicts nature of also acting in vain, and this is inconsistent with our earlier statement.[3] Now the case of the seed of the elm and the willow is in dispute, but the wild fig is conceded to generate seed, and seed[4] that

4.3

we get to these the terms male and female are used not in the proper sense but because of a certain similarity and analogy, for there is in the animals a slight difference of the sort.) "For in plants too some trees of the same kind are fruit-bearing, but others do not bear fruit themselves but contribute to the concoction of fruit in those which do, as happens with the fig and the wild fig"; iii. 5 (755 b 7–11): "There are some who assert that all fish are female except the cartilaginous ones. They are wrong. For they fancy that the females among them differ from those reputed to be male just as in those plants where one bears fruit and the other has none, as olive and wild olive and fig and wild fig [olive and fig are feminine in Greek, wild olive and wild fig masculine]." In the *History of Animals*, v. 32 (557 b 25, 29) in the discussion of caprification he speaks of the *erineoi* of the wild fig in which the insects occur and of the *erina* attached to the cultivated fig. He evidently did not regard these *erineoi* and *erina* as fruit, since they were inedible (*erina* is also used of the unripe figs of the cultivated tree).

THEOPHRASTUS

ἄπεπτον εἰς τὴν ἡμετέραν τροφήν, γεννᾷ δὲ καὶ τὸ τοῦ θύμου[1] ἄνθος (καὶ ἄλλων) ἃ πρὸς τὴν ἡμετέραν αἴσθησιν ἀφανῆ, τὸ δ' ἄνθος μόνον φανερόν.

ἀλλὰ τὰ μὲν τῶν δένδρων (ἢ τινῶν γε[2]) καὶ ἄλλας ἴσως ἀπορίας ἔχει· καὶ γὰρ τὸ[3] περὶ τοὺς φοίνικας ἄτοπον[4] καὶ λόγου δεόμενον, καὶ τὸ οὕτως ἀμενηνά τινων εἶναι τὰ σπέρματα, καθάπερ 4.4 καὶ τῆς κυπαρίττου. τὰ γὰρ τοιαῦτα δίδωσίν τινα ἔννοιαν καὶ ὑπὲρ τοῦ μὴ ἔχειν ἔνια τροφὴν πρὸς ἑαυτοῖς· πλὴν ἴσως ἄλλοις ἄλλη καὶ ὕλη καὶ τροφή, καὶ ἱκανόν τισιν ἂν ἔχῃ τὸ διατηρῆσον.

ὑπὲρ δὲ τῶν σιτηρῶν, καὶ ὅλως τῶν ἐπετείων, αἱ μὲν τοιαῦται διαφοραὶ ῥᾴους, οἷον αἱ κατὰ τὰς

---

[1] u : θυμοῦ U.
[2] Scaliger (τινῶν ἔτι Wimmer) : ἢ τινῶν τε U.
[3] aP : τὰ U N.
[4] U^r N aP : ἀτοπόπον U^{ar}.

---

[1] *Cf. HP* 3 1. 3 (the elm is said by some to be propagated from fruit carried by the wind): "This appears to be similar to what happens with certain undershrubs and herbaceous plants: although they have no visible seed, but instead some have only a kind of down, others only a flower (as thyme), they grow from these."
[2] *Cf. HP* 1 11. 3 (of seeds): "Some seeds are simply a

is unconcocted so far as human consumption is concerned. (And both the flowers of thyme and of other plants generate seeds that are invisible to men, the flower alone being seen.)[1]

But the seeds of trees, or at least of some trees, perhaps present other difficulties as well. For not only is the case of the date-palm[2] odd and in need of explanation, but there is also the fact that in some the seeds are so tiny, as in the cypress.[3] For such cases suggest the notion that some seeds contain no food within their enclosure; except perhaps that both matter and food are different for different seeds, and that it is enough for some to have protection until they sprout.

4.4

Seed Crops and Annuals: The Greatest Difficulty, Transmutation

Passing to cereals and annuals in general we find that the differences of this sort involve less difficulty, namely the differences in sprouting and

---

stone or stonelike at least and as it were dry ... most obviously those of the date-palm, since the seed does not even contain a hollow but is solid throughout. Still this seed too evidently possesses some fluid and heat, as was said [at *HP* 1 11. 1 (the seed) "... possesses in itself native fluid and heat ...']." But Theophrastus may be referring to the part played by the male flower (*cf. CP* 2 9. 15) or to the production of the fruit in a spathe (*CP* 1 20. 2).

[3] *Cf. CP* 1 5. 4.

THEOPHRASTUS

βλαστήσεις καὶ τελειώσεις καὶ τὰ λοιπὰ πάθη τ
4.5 συμβαίνοντα τοῖς τοιούτοις· περὶ δὲ τῆς ἐξαλλα
γῆς εἰς ἕτερον γένος, ὥσπερ ἐκ πυρῶν <καὶ κρι
θῶν>[1] εἰς αἴρας (καὶ εἰ δὴ πάλιν τῶν αἰρῶν εἰ
πυρούς) καὶ τῶν ζειῶν εἰς βρόμον, ἄτοπον αὐτ
τε τῷ συμβαίνοντι καὶ τῷ ἰδίῳ· μόνα γὰρ δὴ ταῦ
τα μεταβάλλει τῶν σπερμάτων φυσικῶς[2] (ὁ γὰ
ἐκ τῆς τίφης καὶ τῆς ζειᾶς πυρὸς παρασκευῇ
πως καὶ τέχνῃ, καθάπερ οἱ τὰ σπέρματα προβρέ
χοντες εἰς τὴν γλύκανσιν).

4.6 ἄτοπον δὲ πρὸς τῷ μόνα καὶ ὅτι ἰσχυρότερ
δοκοῦντα τῶν χεδροπῶν εἶναι (τὸ γὰρ ἀσθενὲς
εὐφθαρτότερον, φθορὰ δέ τις ἡ ἔκστασις), κα
μᾶλλον πυρόν,[4] ᾗ[5] κριθῆς[6] ἰσχυρότερον. ἔτι δ

---

[1] ego. [2] Scaliger : φύσις ὣς U.
[3] u : -ειή U.
[4] ego : πυρὸς U N aP : πυροῦ u.
[5] ego (εἰ Wimmer) : ἢ U. [6] U N : κριθὴ aP.

---

[1] *Cf. HP* 8 7. 1: "Now of other seed crops none change
naturally into another by a loss of identity, but they say
that wheat and barley change over to darnel, and wheat
does so more ... Now this mutation is peculiar to these
and also to flax ..."; *HP* 8 8. 3: "No seed crops changes over
from one whole kind into another except one seeded wheat
and rice-wheat ... and darnel coming from the natural loss
of identity of wheat and barley ..."

maturing and in all but one of the effects to which such plants are liable; their mutation however from one kind to another, as from wheat and barley to darnel[1] (and again, if it happens, from darnel back to wheat)[2] and from rice-wheat to oats[3] is strange both in the occurrence itself and in its restriction to these, for these are the only seed crops to change by their own nature, since the wheat that comes from one seeded wheat and from rice-wheat[4] makes the change because of a certain act of human planning and art (just as when seeds are soaked to make the fruit sweet).[5]

4.5

It is odd that besides being the only plants to change they do so although they are regarded as stronger than legumes (for the weak plant perishes more readily than the strong, and to depart from one's nature is a kind of perishing); and the change of wheat is the odder[6] insofar as wheat is stronger than barley. Another oddity is that change else-

4.6

---

[2] *Cf. CP* 5 3. 7.   [3] Referred to at *CP* 4 5. 1, 2.

[4] *Cf. HP* 2 4. 1: "Among other plants (*sc.* than trees) bergamot mint is held to change to green mint ... and wheat to darnel. Now in trees these changes occur of their own accord, supposing that they do occur. But the changes in annuals are through human intervention: thus one seeded wheat and rice-wheat change to wheat if sown after being bruised in a mortar ..."

[5] *Cf. CP* 2 14. 3 with note *c*.

[6] Wheat has a greater tendency than barley to change to darnel: *HP* 8 7. 1, cited on *CP* 4 4. 5, note 1.

τὰ μὲν ἰσχυρότερα εἰς τὸ ἀσθενέστερον μεταβάλ-
λει·[1] ἡ δὲ αἶρα καὶ πυροῦ καὶ κριθῆς ἰσχυρότερον,[2] ὥσθ' ἅμα συμβαίνει καὶ τὸ παρὰ φύσιν τοῦ
κατὰ φύσιν ἰσχυρότερον.

4.7 ἡ δὲ διαφθορὰ καὶ ἡ μεταβολὴ δυοῖν θάτερον· ἢ
ἐν τοῖς σπέρμασιν, ἢ ἐν τῇ χλόῃ. σπέρματος μὲν
οὖν γαλάκτωσις, σῆψις, ὅλως διάχυσις, ὧν οὐδὲν
φύσιμον· ἡ[3] δὲ χλόη ῥιζωθέντων ἤδη, τὰς δὲ
ῥίζας μεταβάλλειν ἄτοπον.

ὅθεν δὴ τοῖς τοιούτοις παραπεπεισμένοι[4] τινὲς
ὡς ἀλόγοις, ὅλως οὐδὲ γίνεσθαί φασιν τὴν ἔκστασιν, ἀλλ' ἐπομβρίαις[5] φύεσθαι καὶ συνίστασθαι
τὴν αἶραν, ὃ καὶ καθ' αὑτὴν φαίνεται ποιοῦσα, μὴ
σπαρέντων πυρῶν ἢ κριθῶν, ἐν ταῖς ἐπομβροτά-
ταις χώραις.

4.8 αἱ μὲν οὖν ἐναντιώσεις αὗται δοκοῦσιν ἐλέγχε-

---

[1] U : -ειν Wimmer.
[2] aP : -ότερος U N.
[3] Schneider : φύσει· μόνη U.
[4] Heinsius : περι- U.
[5] Schneider : ὑπομβρίας U : ὑπὸ ὀμβρίας u.

where is of stronger to weaker; but darnel is stronger than either wheat or barley, so that we are also faced with this result: that the plant produced unnaturally is stronger than the one produced naturally.

The corruption and change must occur in one of two ways, either in the seed or in the blade. Now such a change in the seed is either by its becoming milky[1] or by its decomposing, in a word by deliquescence, and none of these allows sprouting. A plant in blade, on the other hand, has already struck root, and it is odd that the roots should change. 4.7

Hence some persons, beguiled by this appearance of unreason, assert that no transmutation takes place whatever, and that darnel instead is made to grow and form by rainy weather, a thing which it is also seen to do by itself in the rainiest districts when no wheat or barley has been sown.

*Seed Crops: Transmutation;
Answer to the Objections (1) From the Facts*

Now these objections are considered to be refuted 4.8

---

[1] *Cf. CP* 3 23. 1; *CP* 4 4. 9, and *HP* 8 6. 1: "It is held ... that the worst sowing is that in half-wet ground, for the seeds perish and turn milky." Such milkiness is especially evident in germinating seeds of cereals with starchy endosperms.

THEOPHRASTUS

σθαι τοῖς ἔργοις (πολλοὶ γάρ, ὥς φασι, σπείραντες πυροὺς ἢ κριθὰς ἐθέρισαν αἴρας)·

τὴν δὲ ἔκστασιν[1] καὶ τὴν μεταβολὴν θείημεν ἂν ἀμφοτέρως[2] γίνεσθαι, καὶ τοῦ σπέρματος ἀλλοιουμένου, καὶ τῶν ῥιζῶν (ἡ γὰρ ἐν τῇ χλόῃ μεταβολὴ δι' ἐκείνας), συμφυεῖς δ' οὔσας οὐκ ἄλογον συμπάσχειν. ἐπεὶ τό γε σπέρμα διαφθαρὲν οὐ φύσιμον ὅλως ἦν (τοῦτο μὲν κοινόν· οὐδὲ γὰρ ῥίζαι φθαρεῖσαι τρέφοιεν ἄν).

4.9 τὴν δὲ τοῦ σπέρματος φθορὰν οὐδεμίαν τῶν εἰρημένων ὑποληπτέον, σῆψιν ἢ γαλάκτωσιν, ἀλλ' ἑτέραν, ἣ γίνεται διὰ πλῆθος τῆς τροφῆς ἐκτηκομένων·[3] αὕτη δὲ τὸ μὲν ὅλον οὐκ ἀπόλλυσιν, μεθίστησι δὲ εἰς ἕτερον, ἐπικρατοῦσά πως τῆς ἀρχῆς. τὸ δὲ συμβαῖνον ὅμοιον τρόπον τινὰ καὶ ἐπὶ τῶν ζῴων, ὡς κατ' ἀναλογίαν, <ἂν>[4] ἢ[5] τὸ θῆλυ κρατήσῃ[6] τοῦ ἄρρενος ἢ καὶ ἔτι μείζων ἐναλλαγὴ

---

[1] Itali : ἐξέτασιν U.
[2] U^c from -ων.
[3] Gaza, Heinsius : ἐκτημένων U.
[4] ego.
[5] aP (*quum* Gaza) : ἢ U : ἢ N.
[6] aP^c : -ει U N P^{ac} (?).

DE CAUSIS PLANTARUM IV

by the facts, since authorities say that many persons have sown wheat or barley and reaped darnel.

(2) *From Arguments for the Change Occurring* (a) *in the Seed*

We would take the position that the departure from one identity and passage to the other occurs in both ways, the seed being altered and so too the roots (the change in the blade being due to these); the roots and seed are of a piece, and it is not unreasonable for the roots to be influenced by what is done to the seed. We saw[1] in any case that the seed when corrupted[2] did not sprout at all (this last point applies to both: the roots too, if corrupted, would not provide food).

We must understand the "corruption" of the seed 4.9 as none of the corruptions mentioned,[1] decomposition or turning milky, but as distinct from them, a kind that arises when the seeds deliquesce from too much food; and this corruption does not destroy the seed wholly but shifts it to something else by somehow mastering the starting-point. The result is similar to what happens in animals too (taking the similarity as one of analogy) when either a female is produced by the female parent's mastering the male or a still greater departure in the direction

---

[1] *CP* 4 4. 7.   [2] And not merely "altered" or passing to another identity.

THEOPHRASTUS

4.10 γένηται πρὸς τὸ[1] παρὰ φύσιν· δεῖ γὰρ δὴ τὴν γῆν ὥσπερ τὸ θῆλυ νοῆσαι, καὶ τὸ ἀνὰ λόγον οὕτω λαμβάνειν. ὅταν οὖν ἐκ ταύτης ἡ τροφὴ πλείων[2] γίνηται[3] διὰ τὰς ἐπομβρίας, τότε τὴν ἔκστασιν συμβαίνειν, οὐ φθείρουσαν μὲν ὅλως τὴν γεννητικὴν δύναμιν, ἐξαλλοιοῦσαν δ' εἰς ἑτέραν. εὐζώου δὲ τῆς φύσεως οὔσης, εὐζωοτέρα[4] γὰρ πολὺ τῆς[5] τῶν ζῴων, εὔλογον καὶ διαμένειν μᾶλλον ταύτην· ἐπεὶ καὶ ἄνευ σπερματικῆς ἀρχῆς αὐτόματα πολλὰ συνίσταται καὶ τῶν ἐλαττόνων καὶ τῶν μειζόνων φυτῶν.

4.11 ὅθεν καὶ τοῦτο ἂν εἴη φανερόν, εἴ τις ἀκολουθήσει τῇ ἀναλογίᾳ· διότι τῇ τῶν σπερμάτων ἀλλοιώσει, καὶ οὐ τῇ τῆς χλόης μεταβολῇ·[6] καὶ γὰρ

---

[1] Schneider : τὴν U.
[2] U^ac : πλείον U^c.   [3] u : γίνεται U.
[4] Gaza (*vivacior*), Schneider : -ότερα U.
[5] Wimmer (*quam* Gaza : <ἢ> τὰ Itali) : τα U.
[6] μεταβολῇ U : ἡ μεταβολή Schneider.

---

[1] *Cf.* Aristotle, *On the Generation of Animals*, iv. 1–4 (766 a 16–770 b 27), where the production of female offspring instead of male, of offspring not resembling the one parent or the other, of offspring not resembling any ancestor, and of offspring not resembling a human being, is explained as due to the failure of the male starting-point to master the female matter.

DE CAUSIS PLANTARUM IV

of the unnatural occurs[1]; for we must think of the earth as the female and take the analogy in this fashion. And so when the food from the earth becomes too abundant because of the rains we must suppose that the loss of identity occurs not by a complete corruption of the generative power but by an alteration of that power to a different one. Since the nature of plants is full of life (for it is much more so than the nature of animals)[2] it is reasonable that this nature should also do better at surviving. (For that matter, even without a seed for a starting-point, many plants, both greater and smaller, are formed spontaneously.)[3]

4.10

Hence our first contention would be established, if one is to follow the analogy: the transmutation is by alteration of the seed and not by mutation of the blade, this being also the way in which the change

4.11

---

[2] *Cf.* Aristotle, *On Length and Shortness of Life*, chap. vi (467 a 10–30): "We must set down the cause why the nature of trees is long-lasting, for they have a cause peculiar to themselves when compared to animals (except insects). Plants have a constantly renewed youth, and hence last long. For there are constantly young shoots while the others grow old. So too with the roots ... Plants resemble insects ... for they live when divided and two and more come from one. The insects get as far as living, but cannot do so long, for they have no organs for it and the starting-point in each of the divided parts is unable to produce any. But that in the plant can, for a plant has a root and stalk everywhere potentially..."

[3] *Cf. CP* 1 1. 2.

THEOPHRASTUS

ἐπὶ τῶν ζῴων οὕτω γίνεται.

καὶ ἅμα δὴ τότ' ἀσθενέστατον, ὅταν ἐν μεταβολῇ τυγχάνῃ τῇ κατὰ τὴν βλάστησιν· ἤδη δ' ἐκβεβλαστηκὸς καὶ ἐρριζωμένον ὥσπερ γέγονεν· τούτου δὴ καὶ μείωσις εἰς αὔξησιν καὶ τροφὴν ἢ καὶ ὅλην φθοράν, οὐκ εἰς μεταβολήν.

ἔτι δὲ καί φασιν οἱ ἔμπειροι φανερὸν εὐθὺς εἶναι τὸ φύλλον ἀνατέλλον τῆς αἴρας, λιπαρώτερον ὂν καὶ ποιωδέστερον καὶ στενότερον, καὶ οὐχ ὕστερον τοιοῦτον γινόμενον.

4.12 ἐκ μὲν οὖν τούτων δόξειεν ἂν <ἐν>[1] τοῖς σπέρμασιν.

οὐ μὴν οὐδὲ θάτερον ἀδύνατον, οὐδ' ἄλογον, ὥστε ἐν τῇ χλόῃ μεταβάλλειν, ἀλλοιουμένων τῶν ῥιζῶν (τοῦτο γὰρ ἀνάγκη συμβαίνειν, εἴπερ ἅπαν ἀπὸ τῆς ἀρχῆς ἄρχεται).

καὶ[2] φαίνεται δὲ τοῦτο καὶ ἐφ' ἑτέρων συμβαίνειν, ὥσπερ τῶν δένδρων ὅσα μεταβάλλει ταῖς

---

[1] ἂν <ἐν> Schneider (ἐν Heinsius) : ἂν U.
[2] ἄρχεται. καὶ ego (ἑτέρας ἕτερον Wimmer) : ἄρχεται δεύτεραι U : ἄρχεται δεύτεραι u N : ἄρχεται δεύτερον P (a omits εἴπερ ... συμβαίνειν).

---

[1] *Cf. HP* 8 7. 1 (of darnel): "... it differs in many ways from wheat and barley, for it has a leaf that is narrow,

occurs in animals.

Again at that very time the plant is weakest, when it is undergoing the change involved in emergence from the seed; whereas when it has already sent out its sprout and struck root it has (as it were) completed the process of becoming, and here, to this later stage, belongs a decrease of powers that affects growth and feeding and even leads to total destruction, but no decrease that leads to transmutation.

Moreover persons experienced in these matters say that the moment the leaf appears it is clearly that of darnel, being oilier, more grass-like and narrower,[1] and does not assume this character later.

From these considerations, then, it would appear that the change is of the seed.

4.12

(b) *in the Roots*

Nevertheless the other supposition too is not impossible or yet unreasonable, that the mutation occurs in the blade with the alteration of the roots (since the roots must change if everything begins from the starting-point).[2]

Again this change from the roots is observed in other plants as well, as in the trees that by receiving

hairy and oily..."

[2] Either the roots or the seed may be taken as the starting-point. For the roots as starting-point *cf. CP* 2 14. 3; 3 9. 14.

THEOPHRASTUS

θεραπείαις τοὺς καρποὺς ἐξ ὀξέων καὶ πικρῶν εἰς γλυκεῖς[1] καὶ ποτίμους, ἢ[2] ἐκ πολυπυρήνων <εἰς>[3] ἀπυρήνους.

4.13 ἡ δ' αἰτία τῆς μεταβολῆς ὁμοίως καὶ ταύτης ἐν πλήθει τροφῆς ὅταν ἐπομβρίαι γένωνται, καὶ ὥς γε δή τινές φασιν μάλιστα τοῦ ἦρος ἐὰν ἥλιοι συνεπιλάμψωσιν,[4] ὡς ἐξ ἀμφοῖν γινομένης τῆς μεταστάσεως. ἐνιαχοῦ γὰρ ἔν τισι τόποις ἑλώδεσιν ὅλως τοῦτο συμβαίνειν[5] καὶ τό γε[6] φύλλον μεταβάλλειν[7] ὥστε πᾶσιν εἶναι φανερόν· ἐὰν δὲ τὰ ἐαρινὰ μὴ γένηται πολὺ γίνεσθαι πυρῶν πλῆθος.

ὥστε τοῦτο μὲν οὐδετέρως ἄλογον. εἰ δὲ καὶ ἀμφοτέρως[8] συμβαίνει τάχ' ἂν οὐδ' ἄτοπον εἴη, διαφορὰν δέ[9] τινα ὡς εἶναι[10] δεῖν[11] καὶ τῶν ἐδαφῶν[12] καὶ τοῦ ἀέρος· ἐπεὶ καὶ σπαρέντων εὐθὺ μεγάλη βλάβη πολυυδρίαν ἐπιγίνεσθαι, συμβαίνει

[1] U^c from -ειας.
[2] N aP : ἡ U : ἢ u.
[3] Schneider.
[4] u : -ψουσιν U.
[5] ego : -νει U.
[6] ego : τε U.
[7] U : -ει u.   [8] Gaza : -οις U.
[9] [δέ] Gaza, Wimmer (δή Schneider).
[10] ego (aliquam esse Gaza : τινα εἶναι, ὡς ἄρα Itali : συναιτίαν εἶναι Wimmer) : τινα ἵνα (after -α a letter or stroke erased U^c?) ως εἶρ (or an uncial η; to ν U^c) αι U : τινα εἶναι ὡς

DE CAUSIS PLANTARUM IV

various kinds of tendance transmute their fruit from acid and bitter to sweet and potable, or from having many stones to having none.[1]

The cause of this mutation[2] to darnel too, the one in the roots, is likewise too much food when there has been rainy weather, and (as some assert) the change is especially apt to occur in spring if the sun comes out afterwards, the passage from the one plant to the other being due to both rain and sun. Thus in some countries in certain marshy districts the change (they say) is general, and the leaf changes over in a way that is evident to everyone; whereas if there is no spring rain the wheat produced is very plentiful.

4.13

The mutation, then, is not unreasonable, whether it occurs in the blade or in the seed. And if it moreover occurs in both ways the mutation would perhaps not even be odd[3] (with the proviso that there should be a certain difference in both the soil and the air). Thus heavy rains immediately after sowing cereals are very harmful,[4] since the conse-

---

[1] *Cf. CP* 3 17. 6–7 and the notes.

[2] As well as of the mutation of the seed (*cf. CP* 4 4. 10).

[3] Additional instances may either increase our wonderment or do away with it: *CP* 2 17. 3.

[4] *Cf. CP* 3 23. 3; *HP* 8 6. 6–7.

ἦραι N : τινα εἶναι ὡς αἶραι aP.

11 [δεῖν] Schneider, Wimmer.

12 Itali (*soli* Gaza) : ἐφάφων U.

THEOPHRASTUS

γὰρ ἀσθενεστέρας εἶναι τὰς ῥίζας, τὸ δ' ἀσθενὲς εὐπαθέστερον.

ἡ μὲν οὖν ἐξαλλαγὴ διὰ τοῦτ' ἂν εἴη.

5.1 τὸ δὲ μόνα ταῦτα πάσχειν, ἔτι[1] δ' ἰσχυρότατα δοκοῦντ' εἶναι (καὶ γὰρ ἡ ζειὰ ἰσχυρόν), μιᾷ τινι λύοιτ' ἂν αἰτίᾳ καὶ τῇ αὐτῇ· διὰ γὰρ τὸ[2] ἰσχυρότατα καὶ πολυρριζότατα εἶναι, πλείστην ἔχοντα τροφὴν καὶ μάλιστα [πάσχειν τὰ] πάσχοντα[3] ὑπομένει, τὰ δ' ἄλλα[4] φθείρεται τελέως. ἔτι δ' ἡ ὥρα καθ' ἣν ὁ σπόρος ἐπομβροτέρα, καὶ πλείω χρόνον ἐν τῇ γῇ γίνεται· τὰ δέ, πρὸς τὸ ἔαρ καὶ διαγελῶντος ἤδη τοῦ ἀέρος, ἐν εὐκρασίᾳ μᾶλλον,

5.2 καὶ εὐθὺς εἰς τὴν βλάστησιν ἡ ἀναδρομή. διὸ καὶ οὐδ' ἕτερον οὐδὲν ἐκ τῆς διαφθορᾶς αὐτῶν, ἄν ποτε διαφθαρῇ, γίνεται. τὸ μὲν γὰρ τὴν αἶραν ἀξιοῦν ἄτοπον, εἰς γὰρ τὸ σύνεγγυς[5] καὶ ὁμογενές

---

[1] U^cc : U^ac omits.

[2] u a (το U^c) : τα U^ac (τὰ N P).

[3] ego (πάσχειν · [τὰ] πάσχοντα δὲ Schneider) : πασχειν τὰ πασχοντα U : πάσχειν τὰ πάσχοντα δὲ u (N aP omit τροφὴν ... πάσχοντα).

[4] u : τὀδ' ἀλὰ U.

[5] U^cc : συγεγγὲς U^ac.

---

[1] *Cf. CP* 4 4. 6.

[2] *Cf. CP* 4 4. 5 (it changes to oats).

[3] *Cf. HP* 8 2. 4: "During the winter the cereals remain

DE CAUSIS PLANTARUM IV

quence is weaker roots, and what is weak is more easily affected.

The departure, then, would be due to this.

The difficulty that these are the only grains affected[1] and the further difficulty that they are so affected although considered strongest (rice-wheat[2] too being strong) can be solved by one and the same reason: because they are strongest and have most roots, these grains endure this containing the greatest quantity of food and being most affected, whereas in the rest corruption is complete. They are moreover sown at a rainier season and spend a longer time underground[3]; the rest on the other hand are sown towards spring, in a better tempered ambience, when the air is turning mild,[4] and proceed at once to send up shoots. This is why no other plant at all arises from the corruption of these, should corruption occur: to expect that darnel should arise from it is absurd, since mutations

5.1

5.2

---

in the blade, but when the season turns mild they send up a stem from their midst . . ."

[4] *Cf. HP* 8 1. 3–4: "Now wheat and barley are sown early [*i.e.*, about the time of the setting of the Pleiades, in September], and further . . . all others that resemble wheat . . . ; and of legumes mainly (one might say) bean and birds' pease . . . ; lupine too is sown early . . . Late sown [*i.e.*, sown at the beginning of spring after the winter solstice] are . . . of legumes those such as lentil, tare and pea; and at both seasons those such as vetch and chickpea, and some also sow bean late . . ."

THEOPHRASTUS

πως αἱ μεταβολαί· τὸ δὲ μηδ᾽ εἰς ἕτερον μηδὲν εἰς ἀσθένειαν ἀνακτέον,[1] ὡς ὅλως φθειρομένων· ἀλλὰ τοῖς ὁμοιοπύροις ἂν εἴη καὶ ὁμοιοκρίθοις μᾶλλον εἰς τὴν αἶραν.[2] τούτων δὲ τὰ μὲν ὅλως οὐχ ὑπομένει διὰ τὴν ἀσθένειαν, ὥσπερ ἡ τίφη· τὰ δὲ εἰς τὸ σύνεγγυς μᾶλλον μεταβάλλει, καθάπερ ἡ ζειὰ πρὸς τὸν βρόμον.

5.3 ὁ δὲ πυρὸς εἰς τίφην οὐ μεταβάλλει καὶ ζειὰν ἐξαμβλούμενος, ὅτι πλείων ἡ τροφὴ καὶ ἰσχυροτέρα,[3] δι᾽ ἣν συμμένει· τοιαύτη δ᾽ οὖσα οὐκ ἂν ἐκθηλύνειεν οὐδέν,[4] εἰς <δὲ>[5] τὸ σφοδρότερον ἀγάγοι καὶ ὅλως ἐκστήσειε τοῦ γένους. ἐπεὶ μᾶλλον ἄν τις εὐλόγως θαυμάσειεν ὅτι οὐκ εἰς τὸν ἄγριον[6] πυρόν, ὥσπερ καὶ ἄλλα.[7] τυγχάνει δὲ καὶ τούτου παραπλησία τις ἡ αἰτία· μετακινεῖ γὰρ ἁπλῶς ἡ φύσις.

---

[1] Schneider : ἀκτεον U.
[2] ego (*in lolium* Gaza : εἰς αἶραν Itali : ἡ μεταβολὴ εἰς αἶραν Schneider) : εἰστ᾽ ἤραν U.
[3] u (ἰ- U^c) M aP : ἰσχυρότερα U^{ac} (ἰ- N).
[4] ego (*sed* Gaza : ἀλλ᾽ Itali) : οὐδὲ U.
[5] ego.   [6] U^c : ἄγιον U^{ac}.
[7] <τὰ> ἄλλα Schneider.

---

[1] To these wheat-like and barley-like grains belong rice-wheat, once-seeded wheat, *ólyra* and haver-grass (*HP* 8 9. 2).   [2] Literally "by being aborted."

are into what comes close and is (in a way) of the same kind; and their failure again to change to any other must be traced to weakness, their corruption being entire. It is rather in wheat-like and barley-like grains[1] that a change to darnel would occur. But of these some are too weak to survive at all, as single-seeded wheat, whereas others change into something closer to themselves, as rice-wheat to oats.

On the other hand wheat does not change by arrested development[2] to one-seeded wheat or rice-wheat because its food is too abundant and too strong for that, and the food is why it retains its identity; food of this character, far from producing any feminality,[3] would shift a plant to greater vigour and lift it entirely out of its kind. In fact one would more reasonably wonder rather why the food does not change it to wild wheat, just as food makes other cultivated plants go wild.[4] Here too the reason is a similar one: the nature of the plant makes the shift absolute.[5]

5.3

[3] *Cf.* Aristotle, *On the Generation of Animals*, iv. 6 (775 a 14–16): "... for the females are weaker and colder in their nature, and we must take feminality to be (as it were) a natural stunting of development."

[4] As in the change of bergamot mint (*CP* 4 5. 6).

[5] And not qualified. Change to wild wheat would be change from wheat with the qualification "cultivated" to wheat with the qualification "wild"; change from wheat to darnel is unqualified and absolute.

THEOPHRASTUS

5.4 ἄτοπον δὲ καὶ λόγου δεόμενον εἰ καὶ τὸ λίνον ἐξαιρεῖται, μεγάλη γὰρ ἡ διάστασις· εἰ μὴ ἄρα καὶ τοῦτο τῆς τροφῆς ὁλκόν, φιλεῖ γοῦν χώραν ἀγαθήν, ὥστ' ἐκ τῆς ὑπερβολῆς ἡ διάστασις.[1]

ὁ δὲ θέρμος, ἰσχυρὸς ὢν καὶ πρωϊσπορούμενος,[2] οὐδὲ εἰς ἓν μεταβάλλει διὰ τὴν ἄγαν ἰσχύν, ἐπικρατεῖ[3] γάρ, δεῖ δὲ τὸ μέλλον μεταβάλλειν μήτ'[4] ἀπαθὲς εἶναι μήτε ἄγαν εὐπαθές· τὸ μὲν γὰρ οὐ μετακινεῖται, τὸ δ' ὅλον εἰς αὗον[5] φθείρεται, καθάπερ ἐλέχθη.

5.5 σκεπτέον δὲ καὶ εἴ τι τῶν ἄλλων σπερμάτων, ἢ τῶν ἀγρίων ἢ τῶν ἡμέρων, δέχεται τὴν τοιαύτην ἀλλοίωσιν.

καὶ περὶ μὲν αἰρῶν[6] ἀρκείτω τὰ εἰρημένα.

τὰς γὰρ ἐν αὐτοῖς τοῖς γένεσιν τῶν πυρῶν μεταβολάς (οἷον[7] <ἂν>[8] ἐκ τοιῶνδε τοιοίδε γίνωνται·[9] καὶ κριθαὶ καὶ τἆλλ' ὁμοίως) οὐκέτι

---

[1] U : ἔκστασις Wimmer (*ista quoque mutatio* Gaza).
[2] ego (πρώϊος σπειρόμενος Scaliger : *mature ... obseratur* Gaza) : πρώϊος ἐξαιρούμενος U.
[3] u : ἐπεὶ κρατεῖ U.
[4] Schneider (*neque* Gaza) : μὴ U.
[5] ego (ὅλως ἀσθενὲς ὂν Wimmer) : ὅλον ὡς αὗον U.
[6] u : αἱρετῶν U.   [7] U^c : οἵων U^{ac}.
[8] ego : <ὅτε> Schneider (*quum* Gaza) : <εἰ>Wimmer.
[9] ego : γίνονται U.

DE CAUSIS PLANTARUM IV

But that flax[1] too should change to darnel is odd and requires explanation, since the disparity between them is wide; unless it is because flax too attracts a great deal of food. Thus flax is fond of good land; and the interval covered would be due to the excess of food.

Lupine, which is strong and sown early,[2] changes to no other plant at all because of its exceptional strength. For the strength makes it prevail, and the plant that is to change must be neither unaffected nor very easily affected, since the unaffected plant undergoes no transmutation, whereas the easily affected plant undergoes total corruption and withers away, as we said.[3]

We must also investigate whether any other grain, either wild or cultivated, admits this sort of change of character.

As for darnel, the preceding discussion must suffice.

As for the changes that occur within the various kinds of wheat themselves—that is, wheat of one sort changing to wheat of another, and so too with barley and the rest[4]—we do not seek an answer, as

5.4

5.5

---

[1] *Cf. HP* 8 7. 1, cited in note 1 on *CP* 4 4. 5.

[2] Like wheat and barley. For its early sowing *cf. HP* 8 1. 3, cited in note 4 on *CP* 4 5. 1.

[3] *CP* 4 5. 1; 4 5. 2.

[4] The change of imported grain to the character of the native variety is meant: for this *cf.* note *a* on *CP* 1 9. 3.

THEOPHRASTUS

ζητοῦμεν, οὐδὲ[1] τοῦτ' ἔχει τὸ θαυμαστόν · ἡ γὰρ χώρα[2] καὶ αἱ τροφαὶ καὶ ὁ ἀήρ[3] (καθάπερ εἴρηται) ποιοῦσι τὰς ἀλλοιώσεις ὁμοίως ζῴων καὶ φυτῶν.

5.6 ὃ καὶ πρὸς τὴν ἔκστασιν ὅλως τοῦ γένους χρῆναί φαμεν μετενεγκεῖν, ποιάν τινα καὶ ποσὴν ποιήσαντας τὴν τροφήν · ἐπεὶ καὶ ἡ[4] τῶν δένδρων, καὶ ὅλως τῶν φυτῶν αἱ ἐν ταῖς ἱστορίαις εἰρημέναι μεταβολαὶ διὰ ταύτας γίνονται τὰς αἰτίας (ὥσπερ ἐν τοῖς πρότερον ἐλέχθη) · τῶν μὲν ἀτροφούντων[5] ἐν ταῖς[6] οἰκείαις τροφαῖς καὶ θεραπείαις, ὥσπερ τὸ σισύμβριον ὅταν εἰς μίνθαν,[7] ἀπόλλυται γὰρ τὸ δριμὺ τῆς ὀσμῆς καὶ οἷον ἀποθηλύνεται, διὰ τὴν τροφὴν δὲ καὶ ἀργίαν ἡ ἀπαγρίωσις · ἔνια δ' ὅλως καὶ ἀπόλλυσιν, ὥσπερ τὴν μίνθαν, καταπνιγομένων τῶν ῥιζῶν ὑφ' ἑαυτῶν.

5.7 ἡ δὲ λεύκη τὸ μὲν ὅλον οὐ πόρρω τῆς αἰγείρου καὶ τῇ ὅλῃ μορφῇ καὶ τοῖς φύλλοις · ἀπογηράσκουσα δὲ ἐξομοιοῦται τῷ καταξηραίνεσθαι καὶ

---

[1] Schneider : οὖν εἰ U.
[2] ego : ὥρα U.
[3] U^c : ἀνὴρ U^{ac}.
[4] [ἡ] a.
[5] U^c (-οῦν- U^{ac}) N P : εὐτροφούντων a.
[6] U : μὴ Wimmer.
[7] N aP : μίθαν U.

we did with the other changes, nor does this occurrence have the quality of the marvellous, for the country and the different food and the air (as we said)[1] produce alterations of character in animals and plants alike. One must transfer this explanation, I say, and apply it to the complete departure of a plant from its kind, showing that the food is of such and such a quality and quantity. Indeed not only the change in trees[2] but also the changes of plants in general recorded in the History of Plants[3] come about for these reasons (as we said earlier),[4] some plants getting a starvation diet when they get their appropriate food and tendance (so when bergamot mint changes to green mint,[5] the pungency of odour being lost and the plant becoming as it were female; and this turning wild[6] is due to the increase of food and neglect of tendance); whereas food even completely destroys some plants, as green mint, since its roots choke themselves.[7]

5.6

White poplar is in any case not far removed from black poplar both in its general shape and in its leaves; and when it grows old it not unreasonably is

5.7

---

[1] *CP* 2 13. 1, 5.
[2] *Cf. CP* 2 13. 1–2 16. 8.
[3] *HP* 2 2. 4–2 4. 4.
[4] *CP* 2 13. 1–2 16. 8.
[5] *Cf. HP* 2 4. 1.   [6] *Cf. CP* 2 16. 5.
[7] Compared to bergamot mint it has deep and numerous roots: *cf. CP* 2 16. 5.

THEOPHRASTUS

μᾶλλον ἀτροφεῖν οὐκ ἀλόγως.[1]

ἀλλὰ[2] περὶ μὲν τούτων ἐν ἄλλοις διὰ πλειόνων εἴρηται, καὶ ὅτι δὲ καὶ περὶ τὰ ζῷα κατὰ μὲν τὰς γενέσεις τοιοῦτόν[3] τι συμβαίνει,[4] τελεισθεισῶν δὲ τῶν νοσσιῶν οὐκέτι (πλὴν εἴ τις τὰς κατὰ τοὺς ὄρνιθας ἀλλοιώσεις ἅμα ταῖς γινομέναις ὥραις[5] λέγει·[6] φαίνονται δὲ αὐταί γε πάθεσιν μᾶλλον ὅμοιαι σωματικοῖς ἢ μεταβολαῖς).

6.1   τῶν δὲ ἄλλων τῶν περὶ τὰ σπέρματα μάλιστ' ἄπορον[7] (εἴπερ ἀληθές) τὸ παρὰ μέρος, καὶ μὴ ἅμα, γεννᾶν ἔνια, καθάπερ ἐπί τε τοῦ αἰγίλωπος λέγεται καὶ τοῦ λωτοῦ καὶ τοῦ βολβοῦ. τοῦτο δ' οἱ μέν φασιν εἶναι ψεῦδος, ἀλλὰ ἀπὸ τῆς ῥίζης βλαστανόντων τῶν ὕστερον, ἔτι[8] ἀπὸ τοῦ σπέρματος ὑπολαμβανόντων·[9] οἱ δ' ὡς ἀληθῶς[10] δια-

---

[1] ego: ἄλογον U.   [2] u aP: ἀλλα U : ἄλλα N.
[3] U^ac from τοιοῦτι.   [4] U^r N aP: -ειν U^ar.
[5] ego: τοις γινομένοις ὥραν U.
[6] U^c: -ειν U^ac.
[7] μάλιστα ἄπορον Wimmer: μαλιστα πόρον U.
[8] U: regerminare anno posteriori: aut Gaza: βλαστάνειν τῷ ὕστερον ἔτει ἢ Schneider.
[9] ego (ὑπολαμβάνουσιν Schneider): ἐναπολαμβανόντων U (dot over first α).
[10] U: -ὲς Schneider.

246

assimilated to it by getting dry and taking less food than before.

But these cases have been treated at length elsewhere,[1] as has the point[2] that in animals too something of the sort (it is true) occurs in the process of generation but no longer occurs when the young have been fully formed (unless one brings up the seasonal alterations in birds[3]; but these last evidently bear a closer resemblance to bodily affections than to mutations).

The Next Greatest Problem: Alternate Germination in Annual Seeds

Of other points concerning annual seeds the most difficult is this (if true): that the germination of some seeds is alternate and not simultaneous, as is said of haver-grass, trefoil and purse-tassels.[4] Some assert that this is false; that the later plants come up from the root, but are supposed by the other side to be produced from the seed. But the

6.1

---

[1] *CP* 2 16. 2–5.
[2] *CP* 2 16. 6–7.
[3] *Cf. CP* 2 16. 6 and note *b*.
[4] *Cf. HP* 7 13. 5: "The following is said to be peculiar to purse-tassels, that it does not come up at the same time from all the seeds, but from one seed the same year, from another the next year, as they say is true of haver-grass and trefoil. This then, if true, is a common character of several plants."

THEOPHRASTUS

τείνονται,[1] σημεῖα φέροντες ἄλλα τε καὶ <ὡς>[2] οἱ ἀπολλύντες τὸν αἰγίλωπα δύ' ἔτη[3] τὸν <ἀγρὸν>[4] ἀργὸν ποιοῦσιν, ὅπως ἀμφότερα τὰ σπέρματα ἐκβλαστήσῃ καὶ ἐπινεμηθῇ τε καὶ ἐκθερισθῇ τὸ ὅλον ὁ καρπός.

6.2 ἔχει δέ τινα καὶ ἄλλως ἀπορίαν ἡ παρ' ἔτος βλάστησις· καὶ εἴπερ, συνεχὲς ὂν καὶ δίκρουν (ὥσπερ φαίνεται), καὶ ἅμα πίπτον, τὸ μὲν ἀλλοιοῦται καὶ διαβλαστάνει, τὸ δ' ἀπαθὲς διαμένει πάντα τὸν ἐνιαυτόν· ὅσῳ γὰρ ἀσθενέστερον τὸ ἔλαττον (ὃ δή φασι διαμένειν), τοσούτῳ καὶ εὐπαθέστερον ἐχρῆν εἶναι καὶ εἰς τὴν διαβλάστησιν καὶ εἰς τὴν ὅλην φθοράν. ἀνάγκη δὲ δῆλον ὅτι καί, ὅταν ἐκβλαστήσῃ, θάτερον χωρίζεσθαι· μὴ γὰρ χωρισθέν, καὶ εἰ μὴ γίνεται συμπαθές, ἀλλ' ἐξαιρουμένων γε τῶν ῥιζῶν συνεξαιροῖτ' ἄν, πολλάκις δὲ τοῦτο δρῶσιν οἱ γεωργοί, καὶ οὔ φασι φθείρειν.

---

[1] Schneider : διάκεινται U.
[2] ego : *quod* Gaza : <ὅτι> Schneider.
[3] ego (*biennio* Gaza : διετῆ Scaliger) : δυετήεις U : δυ' ἔτη · εἰς u.
[4] Schneider.

[1] *Cf. HP* 8 11. 9: "A peculiar feature cited for havergrass in contrast to other cereals is this: that one of the seeds takes a year longer than the other to germinate.

DE CAUSIS PLANTARUM IV

first side insists that it really happens, and comes forward with proofs, among them the practice of farmers who wish to get rid of haver-grass of letting the field lie fallow for two years on end so that both sets of seed may come up and the whole crop be grazed over and cut out.[1]

Even so such sprouting with a year's difference is attended with further difficulty: if the two seeds are of a piece with two segments (as we observe) and the two are dropped at the same time, that the one segment nevertheless is altered and sprouts, whereas the other continues unaffected for that whole year. For the greater the weakness of the smaller segment (the one which they say continues unaffected), the more readily should it be affected not only so as to sprout but also so as to perish totally. Further when the one seed sprouts the other must clearly become detached, since otherwise, even if it is not affected along with its partner, still when the roots are pulled up it would be pulled up along with them, and farmers do this often and say that they do not thereby destroy the plant.

6.2

Hence farmers who wish to extirpate it completely (and it is not easy to extirpate) allow their fields to go unsown for two years, and when the plant comes up let their sheep in repeatedly to graze, until the plants have been destroyed by the grazing, and in this way the plant is completely extirpated. And the procedure also testifies to the alternation in the sprouting of the seeds."

THEOPHRASTUS

6.3 τὸ δὲ διαρκεῖν ἀπαθὲς ἐν τῇ γῇ, μέχρι μὲν τῆ<
οἰκείας ὥρας εἰς τὴν ἔκφυσιν εὔλογόν τε καὶ ἐπ
πολλῶν γινόμενον · τὸ δὲ ὅλον ἐνιαυτὸν ἐπισχεῖ<
ἤδη θαυμασιώτερον. ἀλλὰ μὴν καὶ τὸ μὲν τέλεον
τὸ δὲ ἀτελὲς αὐτῶν εἶναι, τελεοῦσθαι δὲ παρ' ἐνιαυτόν, ἄλογον · οὔτε γὰρ ἀσθενές, ἀλλ' ἰσχυρὸν ὁ
αἰγίλωψ, ὥστε ἐκτελεοῦν δύνασθαι πλείω (κα<
ἅμα δὴ καὶ τὸ ἀσθενέστερον τοῦτο δρᾷ) · ἥ τε αἰ
τελέωσις ἡ κατὰ φύσιν οἰκειοτάτη, τάχα δὲ καὶ
μόνη · τροφὴν γὰρ τότε λαμβάνει τὴν αὐτοῦ, διὰ
δὲ τῆς τροφῆς ἡ αὔξη καὶ τὸ τέλεον, χωρισθὲν
δὲ ἤτοι φθείρεται πάμπαν ἢ ἀναυξὲς καὶ ἀνεπί-
δοτον · τὸ δ' αὖ πολύχιτον[1] [δ'][2] εἶναι καὶ τῷ
βρόμῳ καὶ τῇ ζειᾷ συμβέβηκεν, ὥστ' ἐχρῆν[3] καὶ
ταῦτα διαμένειν ἀπαθῆ.

6.4 τὰ[4] μὲν οὖν ἀντιβαίνοντα[5] σχεδὸν ταῦτ' ἐστίν.
[ἀλλ'][6] οὐ μὴν ἀλλ' εἰ δεῖ λέγειν τιν'[7] αἰτίαν,
ἐκείνην ἄν τις ἴσως μάλιστα εἴποι, τὴν φάσκουσαν

---

[1] Schneider : πολυχίτων U N aP : πολυχίτωνι u.
[2] aP.
[3] Wimmer (ὥστε χρῆν Schneider) : ὥστε χρὴ U.
[4] u : ταῦτα U.
[5] U^r : -μαίν- U^{ar} : -μέν- N aP.
[6] Schneider.
[7] U : τὴν u.

250

DE CAUSIS PLANTARUM IV

That this seed should survive unaffected in the ground until the season appropriate for its sprouting is reasonable and happens with many other plants[1]; but that it should hold off a whole year begins to be astonishing. On the other hand, that of the two seeds one should be perfect, the other imperfect, but perfected a year later, is unreasonable, for haver-grass is not weak, but strong, and so is capable of perfecting more seeds than that (then too even a weaker plant does this); and again the natural way[2] is the most appropriate, and perhaps the only way too, of perfecting the seed, since it then receives its own food,[3] and from the food comes growth and perfection, whereas when the seed is detached it either perishes completely or fails to grow and develop. As for its having a number of coats, this is also true of oats and rice-wheat, and on this showing these seeds too should remain unaffected in the ground.

6.3

These then (I may say) are the points that make difficulty.

6.4

Nevertheless if one is to give some reason for the year's difference one would perhaps incline most to

---

[1] *Cf. HP* 7 1. 7 (of herbaceous plants): "Each seed, if it is robust when it drops, waits for its own season and does not come out before . . ."

[2] That is, while the seed remains attached to the plant.

[3] That is, food that has been worked up for it, and not unprepared food taken directly from the ground.

THEOPHRASTUS

μὴ τελεοῦν ἄμφω πρὸς τὸ¹ φύσιμα² ποιεῖν, ἀλλ' ὅμοιόν τι ξυμβαίνειν³ καὶ τῶν ᾠοτόκων καὶ τῶν σκωληκοτόκων⁴ τισί· τὰ γὰρ [ᾠὰ]⁵ ἀποτικτόμενα τρέφεται καὶ ἐκτελεοῦται, τὰ μὲν ἐν τῷ ὕδατι καὶ τῇ θαλάττῃ, τὰ δὲ ἐν τῇ γῇ καὶ τῷ ἀέρι, καὶ ταῦτα δεχόμενα ζῳοποιεῖ, τὰ δὲ φύσαντα καὶ ἐκτεκόντα ἐξαδυνατεῖ.

6.5 φαίνεται δὲ τοῦτό γε καὶ ἄλλως ἀληθές· ὡς οὐκ εὐθὺς ἀλοαθέντα⁶ τὰ σπέρματα βελτίω τῶν χρονισθέντων, οὐδὲ τὰ νέα τῶν ἕνων (ὥσπερ εἴπο-

¹ Schneider: τὰ U.
² aP: φύσημα U N.
³ u, Schneider: -ει U aP (συμβαίνει N).
⁴ Scaliger: σκληρωτόκων U^ar: σκληροτόκων U^r: σκληροτάτων N aP.
⁵ ego.
⁶ αλοαθεντα U: ἀλοηθέντα Wimmer (cf. CP 4 12. 9 ἠλοημένους U: CP 4 12. 8 ἀπηθαλωμενοις U).

---

¹ Cf. Aristotle, *On the Generation of Animals*, ii. 1 (732 a 25–732 b 7): "Of animals some perfect their young and bring them forth like themselves ..., but others bring forth young that are unarticulated and have not yet received their own shape. Of such animals the blooded ones lay eggs, the non-sanguineous larvae ... Of the ovi-

DE CAUSIS PLANTARUM IV

the following, which asserts that the plant does not perfect both seeds to the point of making them fit for germination, but that what happens is similar to what happens in certain oviparous and vermiparous animals[1]: the eggs and larvae on being brought forth feed and are perfected, some[2] in fresh water and the sea, the rest[3] in the earth and in the air, and taking these substances into themselves produce animals, whereas the parents that generated them and brought them forth are unable to take them so far.

This point about imperfection at least appears in any case true: thus seed is not better right after winnowing than after being kept some time, nor yet are this year's seeds better than last year's (as we

6.5

---

parous animals some put forth an egg that is perfect ... (for the eggs of these after coming out cease to grow), but others bring forth imperfect eggs, as fish and crustacea and cephalopods, for the eggs of these grow after coming out"; *ibid.* iii. 9 (758 b 15–21): "For all larvae on coming forth and acquiring size become as it were an egg ... The cause of this is that the nature (*sc.* of the insect) as it were lays eggs before their time because of its own imperfection, as if the larva were a soft egg still in the process of growing."

[2] The eggs.
[3] The larvae.

THEOPHRASTUS

μεν), ἀλλὰ δεῖ τινα λαβεῖν ἐν ἑαυτοῖς οἷον πέψιν καὶ δύναμιν, ἀποπνεύσαντος τοῦ ἀλλοτρίου. τί οὖν, ἴσως ἄν τις φαίη, κωλύει καὶ ἐπὶ τῶν ἀπορουμένων τοιοῦτόν τι συμβαίνειν ὥστε καὶ τελέωσιν λαμβάνειν καὶ δύναμιν ἐκπεττόμενά πως τῇ θερμότητι τῇ τε ἐν ἑαυτοῖς καὶ τῇ περιεχούσῃ;

6.6 ἢ τοῦτο παράλογον, ὡς μᾶλλον ἐν τῇ γῇ κατακλειομένη ἡ θερμότης ἐκπέττει τὴν τροφήν (ἣν ἐπισπᾶται καὶ δι' αὐτῶν)[1] τῆς[2] ἐν τῷ ἀέρι, καὶ ταῦτα πεφυκότων ἐν τούτῳ τελεοῦσθαι (καὶ ἅμα λαμβανόντων τροφὴν[3] οὐκ ὠμήν, ἀλλὰ προ-
6.7 πεπονημένην ὑπὸ τοῦ καυλοῦ καὶ τῶν ῥιζῶν · ἔτι δὲ τῶν μὲν ᾠῶν μία τις ἡ πέψις καὶ τελείωσις ἁπάντων, εἶθ'[4] ὑπὸ τῶν ἔξω); συμβαίνει γὰρ ὥσπερ τὸ μὲν κατὰ φύσιν, τὸ δὲ παρὰ φύσιν, καὶ ταῦτ'[5] ἰσχύοντος ἔτι τοῦ φυτοῦ καὶ οὐ περικαταλαμβανομένου[6] τῇ ὥρᾳ πρὸς τὴν πέψιν.

πλὴν εἴ τις λέγοι ὅτι προεκπηδᾷ τὰ σπέρματα

---

[1] ego : ἢ ἐπισπᾶται καὶ δι' αὐτῶν U : ἣν ἐπισπᾶται, καίτοι αὐτῶν Schneider : τὴν ἐν τῷ σπέρματι καὶ τελειοῖ Wimmer.
[2] Wimmer (τῶν Schneider) : τὴν U.
[3] <τὴν> τροφὴν Schneider.
[4] aP : εἶθ' U N : ἤ θ' u.
[5] aP : ταῦθ' U N.
[6] Schneider : περιλαμβανομένου U.

DE CAUSIS PLANTARUM IV

said),[1] but they need to acquire within themselves a certain concoction (as it were) and power by the evaporation of the foreign element. Someone perhaps might ask: "Then what prevents the like from also happening with the seeds that are the subject of our present difficulty—that the seed acquires full development and power by being concocted in some fashion not only by its internal heat but also by the heat that surrounds it?"

Or is this highly unreasonable? That when the heat is shut up in the earth it does more to bring the food to full concoction (food moreover that the seeds attract by their own efforts) than does the heat in the air, especially when it is natural for seeds to be perfected in the air (where at the same time they get food that is not raw but previously prepared by the stem and the roots; further in the eggs there is a single concoction and maturing of the whole batch, and only then does concoction by external agencies follow)? For the result is that (as it were) the one concoction occurs naturally, the other unnaturally, and this happens when the parent is still strong and not yet precluded from concocting by the season.[2]

Except that one might urge that the seeds drop

---

[1] *CP* 4 3. 4.
[2] So some late tree fruits (seeds) ripen after removal from the tree after the arrival of the cold weather (*CP* 2 8. 2–3).

THEOPHRASTUS

πρὶν τελειωθῆναι διὰ τὸ τὴν φύσιν ἔχειν ἀσθενῆ. συμβαίνει μὲν δὴ τοῦτο καὶ πολὺ προτερεῖ καὶ τῶν κριθῶν, ὥστε κενὸν[1] ἑστάναι τὸν κάλαμον. τοῦτό τε[2] οὖν εἰς τὴν οὐσίαν ἀνακτέον, καὶ τὸ συνεκτρέφεσθαι[3] δὲ καὶ τελειοῦσθαι τοῖς περιέχουσιν οὐκ ἄλογον. ἐπεὶ τό γε διαμένειν ἀπαθές, ἄλλως τε καὶ ἐν χιτῶσι τοσούτοις, ἧττον θαυμαστόν· πολλὰ[4] γοῦν φαίνεται (καθάπερ ἐλέχθη) καὶ τῶν ἄλλων, μᾶλλον δὲ πάντα, διατηρούμενα πρὸς τὴν οἰκείαν ὥραν. ἐκεῖνο γὰρ ἔτι θαυμασιώτερον, ὅπερ εἴρηται περὶ τῶν τευτλίων, ὅτι οὐ μόνον ἑνὶ μηνὶ[5] καὶ δυοῖν καὶ τρισὶν ὕστερον, ἀλλ' ἐνιαυτῷ διαβλαστάνει τινά, καὶ ταῦτα βρεχομένου καὶ κηπευομένου τοῦ τόπου.

6.8 ἔστι[6] δὲ καὶ τὸ τοῦ λωτοῦ σπέρμα τοιοῦτον (ἢ τό γε περιέχον τὸν παρπόν) διὰ τὸ μὴ εὐχώριστον εἶναι μηδ' ὁμοίως ἀφαδρύνεσθαι.[7]

---

[1] u : καινὸν U : κοινὸν N aP.
[2] U : μὲν Schneider.
[3] U : ἐκτρέφεσθαι Schneider.
[4] Gaza (*permulta*), Schneider : πολλαχοῦ U.
[5] Gaza (*uno mense*), Schneider : κ μηνὶ U^{ac} : μηνὶ U^c (κ crossed out) : ἀμηνὶ N : ἐν μηνὶ aP.
[6] U N : ἔτι aP.
[7] Gaza (*capere crassitudinem*), Heinsius (ἀποδύεσθαι Scaliger) : ἀπαδρύνεσθαι U : ἀποδρύνεσθαι N aP.

before maturing because their nature is weak. Now this in fact happens and the seeds drop much sooner than even in the barley,[1] so that the haulm stands empty. This, then, is to be ascribed to the plant's nature; and it is not unreasonable that part of their feeding, together with their maturing, should fall to the environment. As for the seed's remaining unaffected, especially when enclosed in so many coats, there is less to surprise one here (thus many seeds of other plants too, as we said,[2] or rather all seeds, are observed to keep until their proper season). For another case is still more astonishing, where we said[3] of beets that some come up not only a month or two or three but a whole year later, even when the bed is watered and tended.

The seed of trefoil too is of this description[4] (or rather the envelope of the fruit), since it does not separate easily[5] or acquire the same stoutness in all cases.

6.8

---

[1] *Cf. HP* 8 8. 3: "... haver-grass is held to grow by preference in barley ..." Barley was harvested in the seventh month after sowing, wheat in the 8th: *HP* 7 2. 7.

[2] *CP* 4 6. 3 with note 1.

[3] *CP* 4 3. 2 with note 2.

[4] The seeds come up at different intervals.

[5] Presumably the seed does not separate easily from its envelope.

THEOPHRASTUS

τὸ δὲ τοῦ βολβοῦ καὶ τῷ μεγέθει διάφορον· σκεπτέον δ' ὑπὲρ αὐτοῦ.

εἰ δ' οὖν οὕτως ἔνια δυσφυῆ τῶν σπερμάτων ὥστε[1] πολλοῖς ὕστερον χρόνοις ἀνατέλλειν, δῆλον ὅτι ταῦτα πέττεται καὶ τελειοῦται πρὸς τὴν βλάστησιν ἐν τῇ γῇ, διὸ καὶ αὕτη[2] τις ἂν εἴη συνεργοῦσα πίστις. ἀντιβαίνει δ' αὐτῇ τὸ καὶ πρότερον λεχθέν, ὅτι καὶ τελειοῦταί τινα κατὰ τὴν πρόσφυσιν, ἃ καὶ σπειρόμενα διαβλαστάνει παραχρῆμα.

6.9 μεῖζον δ' ἔτι τῶν[3] παρὰ[4] τὰ σπέρματα καὶ τοὺς καρποὺς ἀπορημάτων[5] (οὐχ ὑποπίπτοντα[6] δὲ ὑπὸ τὴν αἰτίαν ταύτην) τὸ περὶ τὸν λωτὸν συμβαῖνον (τοῦτο δέ ἐστι δένδρον παρόμοιον ἔχον τὸν καρπὸν τῷ λωτῷ) καὶ τὸ περὶ τὸν τιθύμαλλον [τὸ περὶ][7] τὸν μυρτίτην καλούμενον (τοῦτο δ' οὐ

---

[1] Gaza (*ut*), Scaliger (τε ὥστε a) : ὥσπερ U N P.
[2] Gaza (*id*), Scaliger : αὐτῃ U : αὐτὴ u (-ή aP) : αὐτῇ N.
[3] Itali : τοῦ U.
[4] U : περὶ Gaza (*de*), Heinsius.
[5] U : ἀπορήματος u.
[6] N : ὑποπίπτοντα U : ὑποπίπτοντος u : ὑποπίπτον aP.
[7] u : καὶ τὸ περὶ aP.

DE CAUSIS PLANTARUM IV

In purse-tassels[1] the seed also varies in its size. We must investigate it.

At all events if some seeds are so slow to sprout that they come up long after they are dropped, it is evident that they are concocted and matured for germination in the ground, and so we have here a piece of corroborative evidence. But opposed to this evidence is the point mentioned before[2]: that some seeds are perfected when still attached, and these come out right after sowing.

A Greater Difficulty for Consideration

A still greater difficulty than those depending on variations in the annual seed and fruit (but these new cases do not fall under this cause)[3] is what happens in the *lotos*[4] (this is a tree with fruit resembling that of trefoil)[5] and with the so-called "myrtle"

6.9

---

[1] *Cf. HP* 7 13. 5: "The following feature is cited as peculiar to purse-tassels, that it does not come up simultaneously from all the seeds, but from one seed in the same year, from another in the next, as is said of haver-grass and trefoil."

[2] *CP* 4 6. 1 (*cf. HP* 7 13. 5, just cited).

[3] Concoction of the seed (fruit) in the ground.

[4] Perhaps the Libyan *Zizyphus lotus*. But the name *lotos* is also used of the nettle-tree (*Celtis australis*).

[5] *Lotos* in Greek.

THEOPHRASTUS

δένδρον, ἀλλὰ θάμνος ἐπιγειόκαυλος)· ἀμφότερα γὰρ ταυτά φασι παρὰ μέρος φέρειν ἐξ ἑκατέρων τῶν βλαστῶν, ὥστ' ἀνάγκη τὸν μερισμὸν τῶν δυνάμεων ἐν ταῖς ῥίζαις ποιεῖν.

ἀλλὰ περὶ μὲν τούτων ἐπισκεπτέον.[1]

7.1 τῶν δὲ σιτηρῶν σπερμάτων εἴ τινες διαφοραὶ κατὰ τὴν ἔκφυσιν ἢ τοὺς καρποὺς ἢ τὴν ἄλλην οὐσίαν δεῖ πειρᾶσθαι λέγειν.

οἷον ὅτι πυρὸς μὲν καὶ κριθὴ καὶ τὰ ὅμοια πρωϊσπορεῖται, τὰ[2] χέδροπα δὲ πρὸς τὸ ἔαρ, πλὴν κυάμου καὶ θέρμου. τοῦτο γὰρ ποιοῦσιν ὅτι τὰ μὲν σιτώδη ῥιζωθῆναι δεῖ, παρὰ[3] τὸ[4] μονόκαυλα[5] γίνεσθαι, τῆς ἀναφορᾶς εὐθὺς ἄνω γινομένης· ῥιζοῦται δὲ κατεχόμενα καὶ πιλούμενα τοῖς χειμῶσιν, ὥστε πολλὰς ἀρχὰς λαμβάνειν τῆς

---

[1] u : ἐπισκεπτειν U.
[2] Schneider (mature seruntur Gaza) : πρῷρεῖ (προρεῖ Uʳ). τα δὲ (δὲ crossed out Uᶜ) U.
[3] ego : πρὸς U.
[4] τὸ <μὴ> Schneider.
[5] MN² : μονόκωλα U (dot over ω) N¹ aP.

---

[1] Cf. HP 9 11. 9: "The so-called 'myrtle' spurge sends out branches on the ground about a span long, and these do

DE CAUSIS PLANTARUM IV

spurge[1] (this is not a tree but a shrub with a stem running on the ground), for both plants are said to bear alternately from two sets of shoots, so that the plants must make this apportionment of their powers in the roots.

But these are matters that we must investigate.

Cereals and Legumes: The Easier Differences[2]

Regarding the seeds of cereals we must endeavour to treat any differences they have in sprouting or in fruit or in the rest of their nature. 7.1

So for instance the difference that wheat and barley and the like are sown early,[3] whereas legumes (bean and lupine excepted) are sown towards spring.[4] This is done because cereals need to get rooted because they become single-stemmed when their growth is directed upward from the start; and they get rooted by being held back and pushed down by the cold weather, with the result that they acquire many starting-points for their upper

---

not bear the fruit at the same time but in alternate years, some bearing now and some next year, although they grow from the same root."

[2] A subject broached at *CP* 4 4. 4, second paragraph.

[3] In autumn: *cf.* note 4 on *CP* 4 5. 1.

[4] *Cf. HP* 8 1. 3–4, cited in note 4 on *CP* 4 5. 1.

βλάστης.

7.2 τὰ δὲ χεδροπά, μονόρριζά τε καὶ ἰσχυρόρριζα, καὶ εὐθὺ τὴν¹ ὁρμὴν ἄνω ποιούμενα, προσφόρως ἔχει τῇ ὥρᾳ, τὸ γὰρ πολὺ τῆς φύσεως ἐν τῷ ἀέρι. τὸν δὲ κύαμον πρωϊσποροῦσιν (ὥσπερ εἴρηται) διὰ τὴν ἀσθένειαν, ὅπως ῥιζωθεὶς² ἐν ταῖς εὐδίαις ἀντέχῃ τοῖς χειμῶσι·³ τὸν δ' αὖ θέρμον εὐθὺς ἀπὸ τῆς ἅλω, διότι μὴ καταβληθεὶς ἔτι θερμῆς οὔσης τῆς γῆς κακοβλαστὴς γίνεται·⁴ τούτου δ' αἴτιον ὅτι ἀτέραμον φύσει καὶ ὥσπερ ἄπεπτον· δηλοῖ δ'

7.3 ἡ πικρότης διότι πολλῆς δεῖται θερμότητος. ἐὰν οὖν προσλάβῃ τὴν ἐκ τῆς χώρας, δύναται κατεργασθῆναι καὶ εὐβλαστεῖν· εἰ δὲ μή, κακοβλαστεῖ,⁵ καὶ παρόμοιον τὸ ξυμβαῖνον ὥσπερ εἴ τις ὀλίγῳ πυρὶ τὸ ἰσχυρότατον ἕψειν ἢ⁶ ὀπτᾶν ἐπιχειροίη. διὰ τοῦτο δ' ἔοικεν οὐδὲ τὴν ἀγαθὴν χώραν

---

¹ ego : εὐθὺς τὴν aP : εὐθετην U (-ην u N).
² u aPᶜ (-ῆς N) : ῥιζωθῆι U (-ῇ P ᵃᶜ?).
³ u N aP : χιτῶσι U.
⁴ U : γίνηται u.
⁵ u : κακοβλαστη U.
⁶ ἕψειν ἢ ego : ἐκεῖνο εἰ Uᵃʳ : ἐκεῖνο Uʳ N aP : κρέας Wimmer.

growth.

Legumes on the other hand, with their single strong root, and their impetus of growth at once directed upward, are well suited to spring sowing, since the greater part of their nature is above ground. But bean (as we said)[1] is sown early because of its weakness, to let it get rooted in the fine weather and so withstand the cold[2]; lupine again is sown straight from the threshing floor,[3] since unless it is sown when the ground is still warm it sprouts poorly. The reason is that the plant is by nature refractory to concoction and (so to say) not concocted at all[4]; and the bitter taste shows that a great deal of heat is required. Now if it gets the additional heat from the ground, the seed can be worked up and sprout well; but otherwise it sprouts poorly and what happens is like attempting to boil or roast with too little fire a piece of food that needs a great deal of cooking.[5] For this reason, it appears,

7.2

7.3

---

[1] *CP* 3 24. 3, 4; *cf. CP* 4 7. 1.

[2] *Cf. HP* 8 1. 3, cited in note 1 on *CP* 3 24. 3.

[3] *Cf. HP* 8 1. 3: "... lupine is also sown early, for we are told to sow it straight from the threshing floor"; *HP* 8 11. 8: "... for lupine, in spite of its great strength, if not sown straight from the threshing floor, sprouts poorly, as we said ..."

[4] *Cf. CP* 4 1. 7.

[5] For this *mólysis* or "half-cooking" *cf. CP* 2 15. 2 with the notes.

THEOPHRASTUS

φιλεῖν, ἀλλὰ τὴν ὕφαμμον·[1] εὐθερμαντοτέρα[2] γὰρ αὕτη καὶ οἷον εὐκατεργαστοτέρα. καὶ κρύπτεσθαι <δ'>[3] οὐκ ἐθέλει κατὰ βάθους, ἀλλ' ἐπιπολῆς, ὅπως μᾶλλον ἀπολαύῃ τῆς θερμότητος, καὶ ποτιμώτερος· καὶ μετέωρος ὢν ἐκβλαστάνει καὶ καθίησι[4] τὴν ῥίζαν.

7.4 καὶ τὰ μὲν τοιαῦτα σχεδὸν ὥσπερ γεωργικά· τὰ δὲ αὖ τῆς[5] φύσεως, οἷον ὅτι τῶν μὲν χεδροπῶν ἐκ τοῦ αὐτοῦ μέρους ἥ τε ῥίζα καὶ ὁ καρπὸς[6] ἐκβλαστάνει, καὶ τὸ μὲν εὐθὺς ἄνω, τὸ δὲ κάτω νεύει καὶ ὁρμᾷ· πυροῦ δὲ καὶ κριθῆς καὶ ὅλως τῶν σιτωδῶν, ἐξ ἑκατέρων τῶν ἄκρων, ἀπὸ μὲν τοῦ παχέος <ᾗ>[7] καὶ[7] πρὸς τῷ[8] στάχυϊ προσπέφυκεν, ἡ ῥίζα, ἀπὸ [δε][3] τοῦ λεπτοῦ <δὲ>[3] ὅθεν ὁ ἀθήρ,[9] ὁ καυλός, ὡς ἂν[10] ἑκάτερον ἀνὰ λόγον,[11] τὸ μὲν ἀπὸ

---

[1] U (cf. HP 8 11. 8): ἔφαμμον Schneider.
[2] U^r N aP: -τέραι U^{ar}.
[3] ego.
[4] Schneider: καθιστησι U.
[5] u: αὐτῆς U.
[6] U: καυλὸς Itali.
[7] ego (ᾗ Schneider: quod Gaza): καὶ U.
[8] τῷ u N aP: το U.
[9] Gaza (arista), Scaliger: ἀὴρ U.
[10] U: ὂν Schneider.
[11] Wimmer: ἀνάλογον U.

it also does not like good soil but prefers sandy,[1] for sandy soil is more easily heated and (so to say) digested.[2] Again, it does not like to be covered deep but to be near the surface,[3] so that it may profit better from the heat, and it is then more palatable; even when the seed fails to hit the ground it sprouts and sends down its root.[4]

Such matters as these belong (one may say) to the art of husbandry. Others belong to the nature of the plants, for instance that in legumes the root and fruiting-stem come out of the same side of the seed, the stem at once turning and moving upwards, the root downwards; whereas in wheat, barley and cereals in general the one part comes from the one extremity, the other from the other: the root from the thick end where the seed is attached to the ear, the stem from the thin end from which rises the awn, each coming from the extremity to which it

7.4

---

[1] *Cf. HP* 8 11. 8 (of lupine): "And it seeks by preference sandy and poor soil, and is in general unwilling to grow in tilled ground."

[2] "Working up" or "digestion" involves the application of heat; so what is more easily heated is also more easily digested.

[3] *Cf. HP* 8 11. 8 (of lupine): "... it does not like to be covered by the ground, which is why it is sown without any ploughing immediately preceding."

[4] *Cf. HP* 1 7. 3, 8 11. 8, cited in note *b* on *CP* 2 17. 7. Thus suspended it gets more heat from the sun.

THEOPHRASTUS

τοῦ ἄνω, τὸ δὲ ἀπὸ τοῦ κάτω, καὶ γὰρ ἡ ῥίζα καὶ ὁ καυλὸς οὕτω.

7.5 τὴν <δ'>[1] αἰτίαν εὐθὺς ἐν τῇ φύσει ληπτέον· ὅτι τὰ μὲν χεδροπά, δίθυρα[2] ὄντα, μίαν τινὰ καὶ

---

[1] Wimmer : τὴν U N P : καὶ τὴν a (no punctuation precedes in the MSS.).

[2] Gaza (*bipartito et compactili ... corpore* [from *HP* 8 2. 2]), Itali : αἴθυρα U : ἄθυρα u Nᶜ (from ἀθύρα) aP.

---

[1] This is a correction of Aristotle, who did not differentiate in this matter between leguminous seeds and the rest. *Cf.* Aristotle, *On Youth and Age*, 3 (468 b 18–23): "For the generation from the seeds arises in all plants from the middle of the seed. For since all seeds are two-valved, they cling to the plant at the place where the valves grow together, and this is the mid-place between the two parts, for the stem and root of plants grow from here, and the starting-point is the middle place between the two"; *cf.* also *On the Generation of Animals*, iii. 2 (752 a 18–23): "Things are the same way (*sc.* as with eggs) in the seeds of plants; for the starting-point of the seed is attached in some cases to the twigs, in others to the husks, and in others to the pericarpia. This is evident in legumes; for the point where the two valves of bean and such seeds are connected is the point where the seed is attached, and here is the starting-point of the seed."

With these passages and *CP* 4 7. 4–7 *cf. HP* 8 2. 1–2 (of seed-crops):

"Some seeds send both root and leaf from the same extremity, but others send the one from the one extremity and the other from the other. Now wheat, barley, one-seeded wheat and in general all cereals send one from

DE CAUSIS PLANTARUM IV

corresponds, the one from the upper extremity, the other from the lower, root and stem having these positions in the plant.[1]

The cause is to be found directly in their nature. 7.5 Leguminous seeds, on the one hand, since they are two-valved, have (one may say) one and the same

each extremity, just as (ὥσπερ Scaliger : πάντα U) the seeds grow on the ear, sending the root from the lower, thick, extremity and the shoot from the upper extremity, and the product of both becomes a single continuous unit. But bean and the other legumes do not do so. Instead they send root and stem from the same extremity, where the seed is attached to the pod, and where they have a sort of starting-point plain to the eye (in some it actually looks like a penis, as in bean, chickpea and especially lupine); for from this part in the various legumes the root moves downward and the leaf and stem move upward.

"In this way legumes and cereals differ. But in another way they are alike, in that all of them send out the root at the place of attachment to the pod or the ear, and not the other way round as in some seeds of trees, for instance almonds, filberts, acorns and the like. In all seeds the root comes out a little earlier than the stem. It happens in cereals at least that the shoot first sprouts inside the seed itself, and as the shoot grows the seed splits open (for all these seeds too are in a way, as in legumes, of two parts, legumes being visibly all two-valved and composite), whereas the root is sent outside directly. But in leguminous seeds this (*sc.* the previous sprouting of the shoot within the seed, with consequent splitting) does not occur, because the valves are separate, although the root comes out a little earlier."

THEOPHRASTUS

τὴν αὐτὴν <ἀρχὴν>[1] ἀμφοῖν ἔχει κατὰ τὸ ἄκρον
(ὃ καὶ φαίνεται προσπεφυκὸς ὥσπερ ἔξωθεν), ᾗ[2]
καὶ συνάπτουσι πρὸς τὸν λοβόν · ἅτε γὰρ ἐναγγειόσπερμα ὄντα ταύτην μὲν ἐξ ἀνάγκης ἔχει, τρέφεσθαι γὰρ ἄλλως οὐκ ἦν · τὴν δέ, ἐκ τοῦ αὐτοῦ[3]
μέρους, ὃ δυνατόν,[4] συνεχῆ τ᾿ ἀλλήλοις ὄντα καὶ
περιεχόμενα τῷ λοπῷ ·[5] τὸ δ᾿ εἶναι τοιαῦτα τῆς
οὐσίας.

7.6 ὥστε τούτῳ μὲν εἰκότως ἡ φύσις εἰς ταὐτὸ τὰς
ἀρχὰς ἔθηκεν · τῶν δὲ σιτωδῶν, καθ᾿ αὑτά τε πεφυκότων καὶ[6] περιεχομένων, ἔτι δ᾿ ὥσπερ εὐθέων
πως ὄντων τὴν μορφήν, ἐχώρισεν[7] τὴν ἔκφυσιν,
ἀπὸ μιᾶς μὲν ἀρχῆς ἀμφοτέρας ποιήσασα, συν-

---

[1] Gaza (*initium*), Heinsius.
[2] ᾗ u : ᾗ U : ᾗ N : ὂν aP.
[3] U : *altera . . . parte* Gaza : μὴ ἐκ τοῦ αὐτοῦ Scaliger : ἐκ τοῦ ἄλλου Heinsius : ἐκ τοῦ ἑτέρου Schneider.
[4] ego (ἀδύνατον Wimmer) : ὁ δυνατὸν U : οὐ δυνατὸν u.
[5] ego : λόγωι U : λοβῶι u.
[6] καὶ <μὴ> Schneider.
[7] U^r from ἐχώρησεν.

---

[1] That is, root and stem.
[2] That is, at the extremity shared by the two valves.
[3] That is, on the outside of the extremity.
[4] Food, whether from the parent or from the seed's own root, must enter the seed at the same place.

DE CAUSIS PLANTARUM IV

starting-point for both[1] at their extremity[2] (an extremity that is observed to be attached to the valves as if from outside), at the place where the valves connect with the pod. For since legumes have their seeds in a pod, they necessarily have this one of the two starting-points here[3] (for the seed could not otherwise be fed)[4]; and they have the other starting-point on the same side,[5] which it is possible for them to do,[6] since here the seeds have their valves continuous with one another and are enveloped by the skin. That the seeds should have this character belongs to their nature as legumes.[7]

And so in the leguminous seed the nature of the plant has understandably put the two starting-points in the same place; whereas in cereal seeds, which grow and are enclosed singly,[8] and which furthermore are (as it were) straight in shape, the nature of the plant divided the place of emergence, letting both emergences arise from a starting-point

7.6

---

[5] That is, at the same end of the seed as the connexion with the pod, but inside the skin.

[6] This location is not dictated by necessity (as with the starting-point of the root). It is a possible location since the skin protects the germ of the shoot and the germ itself is within easy distance of the root.

[7] There is therefore no need to explore why legumes have two-valved seeds contained in pods.

[8] That is, they do not grow as two valves enclosed in the same skin.

εχοῦς δέ τινος οὔσης καὶ διηκούσης[1] ἐφ' ἑκάτερον τῶν ἄκρων, ὥσπερ[2] καὶ κατὰ τὴν βλάστησιν φανερὰ διατείνουσα κατὰ τὴν ἐντομήν. τοῦ μὲν οὖν μὴ ὁμοίως ἔχειν ταύτην ὑποληπτέον τὴν αἰτίαν.

7.7 ἐκεῖντο δ' ὁμοίως ἐν ἀμφοῖν ἐστιν, ὥστε τὴν τῆς ῥίζης φύσιν[3] ἀπὸ τοῦ αὐτοῦ καὶ[4] τὴν ἐν τῷ στάχυϊ καὶ τῷ λοβῷ γίνεσθαι πρόσφυσιν, ἀφ' ὧν[5] ἀμφοτέρων αἱ τροφαί, τοῦ μὲν καλάμου[6] καὶ τῆς πρώτης βλάστης ἡ ῥίζα, τῶν δὲ καρπῶν ἡ πρόσφυσις ἥ τε πρὸς τὸν λοβὸν καὶ πρὸς τὸν στάχυν. (σκεπτέον[7] δὲ <τὸ>[8] τῶν καρυωδῶν καὶ βαλανωδῶν.)

καὶ τὰ μὲν περὶ τὴν ἔκφυσιν οὕτως ἄν τις διέλοι.

8.1 τὴν δὲ ταχυβλαστίαν τοῖς χεδροποῖς ἀποδοίη

---

[1] ἀμφοτέρας ... διηκούσης U : ἀμφοτέρων ποιήσασα, συνεχῆ δέ τινα οὖσαν καὶ διήκουσαν Schneider.
[2] U : ἥπερ Gaza, Schneider.
[3] U : ἔκφυσιν Heinsius (*exortus* Gaza).
[4] U : κατὰ Schneider.
[5] U : οὗ Schneider.
[6] Gaza, Itali : βαλσάμου U.
[7] σκεπτέον ego (... Πίπτοι Schneider) : πίπτοι U.
[8] ego.

that is single, it is true, but that is on the other hand a continuous one, and runs through to each of the two extremities, just as at the time of sprouting the starting-point can be seen extending along the groove.[1] To conclude: we must suppose that the reason for the difference in the place of emergence is this.

But in both the cereal and the leguminous seed 7.7
the following feature is equally present: the root-part is produced from the same side as that where the attachment to the ear and to the pod is formed, and it is these places that are the sources of the food supply for both, the root being the source for the haulm of the cereal and the first shoot of the legume, the attachment to the pod and to the ear being the source for the fruit. (We must investigate the case of nuts and acorns.)[2]

These are the distinctions one would make in the matter of growth from the seed.

Seed-Crops: Relative Speed of Sprouting and Maturing in Legumes and Cereals

One would assign the rapid sprouting of 8.1

---

[1] It can be seen because by then the seed has split open.
[2] *Cf. HP* 8 2. 2, cited in note 1 on *CP* 4 7. 4.

THEOPHRASTUS

τις [1] διὰ τὴν ὥραν, ὅτι μαλακωτέρα καὶ γονιμωτέρα τῆς χειμερινῆς· ἀλλ' ἀμφοῖν τούτοιν αἰτιάσαιτο τὴν ἀσθένειαν, δι' ἣν καὶ σπείρονται πρὸς τὸ ἔαρ, καὶ διαβλαστάνουσιν θᾶττον· τὸ γὰρ ἀσθενέστερον (ὥσπερ πολλάκις ἐλέγομεν) εὐπαθέστερον· τάχα δὲ καὶ τῆς πολυκαρπίας (πολυχούστερα γὰρ δὴ τὰ χεδροπά) τὴν αὐτὴν ἢ παραπλησίαν, καὶ τοῦ θᾶττον δὲ ἐκτελεοῦν, καὶ μὴ καρπίζεσθαι τὴν γῆν, ἀλλὰ νειὸν [2] ποιεῖν· ἅπαντα γὰρ ταῦτα ἢ τὰ τοιαῦτα συμβαίνει σχεδὸν (ὥσθ' ἁπλῶς 8.2 εἰπεῖν) διὰ τὴν ἀσθένειαν. καὶ γὰρ ἐκθρέψαι καὶ τελειῶσαι ῥᾴδιον [3] τὸ ἀσθενέστερον, ὡς ἅμα καὶ πλεῖον [4] καὶ τὴν γῆν ἧττον καρπίζεσθαι, καὶ τὴν ἔκφυσιν δὲ τὴν πρώτην ῥᾷον [5] καὶ θᾶττω ποιεῖ-

---

[1] τις <ἂν> Wimmer.
[2] Nc (ἀλλὰ in an erasure) aP : ἀλλ' ανειον U.
[3] U : ῥᾷον Gaza (*facilius*), Wimmer.
[4] U : *ut et fructificetur uberius* Gaza : ὥσθ' ἅμα καὶ πλεῖον (φέρειν Schneider) φορεῖν Itali : ὥστε καὶ πολυκαρπεῖν Wimmer.
[5] Uc (ῥαίον Uac : ῥᾷον N) : ῥάω aP.

---

[1] *Cf. HP* 8 1. 5 (wheat and barley come up on the seventh day, pulses on the fourth or fifth.)
[2] *CP* 2 11. 6; 4 1. 3; 4 3. 6; 4 6. 2.
[3] *Cf.* Aristotle, *On the Generation of Animals*, i. 18 (726 a 9–11) (some things have much seed and are prolific

legumes[1] to the season, because it is milder and more procreative than winter. But one would give their weakness as the reason for both the season and the speed, since it is owing to their weakness that they are sown towards spring and that they come up sooner, for the weaker plant (as we have been saying)[2] is the more easily influenced. Perhaps one would give the same reason or one close to it for their abundant crop[3] as well (legumes having a greater yield than cereals)[4] and for their maturing it sooner, and not exhausting the soil but renewing it[5]; for all these things (or things of the sort) result (one may say, to put it in a word) from weakness. For it is easy both to bring up and to 8.2 mature the weaker seed, which leads to a greater yield accompanied by less exhaustion of the soil and also to easier and quicker production of the first emergence from the seed (except when a seed is

(πολύχοα) because of power, and some because of lack of power). The word πολύχους ("prolific," literally "with many *choes*; a *chous* contained about six pints) implies seed-crops; both the seed sown and the crop harvested consisted of "seeds" which could be measured and compared.

[4] *Cf. HP* 8 3. 4: "In general legumes have more fruit and a greater yield, but the summer seeds millet and sesame yield still more than these ..."

[5] *Cf. HP* 8 7. 2: "Peculiar to chickpea compared to the other legumes is ... that it exhausts the soil and does not renew it ... Of the rest the best at fertilizing the soil is bean ..."

σθαι (πλὴν[1] εἴ τι κωλύεται διὰ τὴν ἰσχὺν τοῦ περιέχοντος, ὥσπερ ὁ κύαμος· αἰτία γὰρ τούτῳ τῆς δυσφυΐας ἡ τοῦ κελύφους παχύτης, ἐὰν δὲ καὶ ἐφύσῃ[2] σπαρέντι, δυσφυέστερον ἔτι, καθάπερ ἐδαφιζομένης τῆς γῆς).

ἐν δὲ τοῖς σιτηροῖς καὶ ταῦθ' ὁμολογούμενα πρός τε τὴν χρονιότητα τῆς πέψεως καὶ πρὸς τὸν καρπισμὸν τῆς γῆς· οἷον ἥ τε πολυρριζία[3] καὶ ἡ γλισχρότης τῶν καρπῶν· βραδέως γὰρ τὰ γλίσχρα καὶ καθαρὰ καὶ ἐν τοῖς δένδροις πεπαίνεται, διὸ καὶ πυροὶ κριθῶν ὀψιαίτεροι καὶ ὀλιγοχούστεροι.[4] καὶ αἱ ῥίζαι δὲ τῶν σιτηρῶν,[5] πολλαὶ καὶ κατὰ βάθος ἰοῦσαι,[6] καρπίζονται μᾶλλον, οὐχ ὥσπερ τῶν γε[7] χεδροπῶν[8] ὅ τε καρπὸς γεωδέ-

---

[1] u : τὴν U.
[2] Gaza (*impluerit*), Itali : ἐκφυσήσῃι U.
[3] aP (πολυριζία u N) : πολυρία U.
[4] aP : ὀλιγοχουστότεροι U (ὀ- u N).
[5] U^cc : δένδρων U^ac (now crossed out).
[6] Schneider : οὖσαι U.
[7] aP : τε U N.
[8] χεδροπῶν <ὧν> Schneider.

---

[1] *Cf. HP* 8 1. 5 (of bean): "... for it is the slowest to come up of all legumes, and if there is fairly prolonged rain after it is sown, it is very slow indeed."

prevented from coming out by the strength of its envelope, as bean, for the reason for the slowness of bean to grow is the thickness of its skin; and if in addition there is rain after it is sown, bean is even slower to grow,[1] since the soil is as it were tamped down).

In cereals on the other hand there are these further reasons[2] that are agreed to make for their hardiness of concoction and their exhaustion of the soil[3]: their many roots and the viscosity of their fruit. For even in trees it takes long to ripen fruit that is to be viscous and rid of earthy admixture[4] (and this is why wheat comes up later than barley[5] and has a smaller yield). So too in cereals the roots are numerous and go deep and thus do more to exhaust the soil, unlike legumes with their earthier

8.3

---

[2] In addition to relative strength compared to legumes.

[3] *Cf. HP* 8 9. 1–3: "Wheat and barley exhaust the soil most ... Of the plants resembling wheat or barley the strongest and most exhausting to the soil is rice-wheat, since it is many-rooted and deep-rooted, and many-haulmed ..., and of the rest oats, for oats too are many-rooted and many-haulmed ... Wild oats too exhaust the soil greatly and are many-rooted and many-haulmed ..."

[4] *Cf. CP* 1 6. 2 and 4 8. 3 (here the fruit of legumes is called earthier than that of cereals).

[5] *Cf. HP* 8 1. 5: "... wheat and barley come up on about the seventh day, barley being a bit ahead ..."

THEOPHRASTUS

στερος καὶ ἡ ῥίζα μία καὶ ἐπιπολῆς <καὶ>[1] ἡ
σπορὰ μανή. πάντα δὲ ταῦτα συνεργεῖ καὶ πρὸς
πλῆθος καρποῦ, καὶ πρὸς κουφότητα τῆς γῆς κα.
τὸ <μὴ>[2] ὁμοίως καρπίζεσθαι, <καὶ>[1] πρὸς
ταχύτητα πέψεως. ἅπασα γὰρ ἡ δύναμις ἄνω
φερομένη, καὶ οὐκ ἀντισπώσης τῆς ῥίζης, πλῆθος
ἀποδίδωσι, καὶ συνεργοῦντος τοῦ ἀέρος ἐκπέττε.
ῥᾳδίως. ἐπεὶ ὧν γε[3] πλείων ἡ ῥίζα καὶ κατὰ
βάθους καὶ οἱ καρποὶ[4] ξυλώδεις, ταχὺ δὲ[5] ἐκκαρ-
πίζεται τὰ ἐδάφη, καθάπερ ὁ ἐρέβινθος · διὸ καὶ
μόνον[6] οὐ[7] ποιεῖ νεόν,[8] οὐδ' ἐστὶ τῆς τυχούσης
χώρας,[9] ἀλλ' ἀγαθῆς.

8.4 ἄλογον δὲ ἐπὶ τούτου φαίνεται τὸ ταχὺ τελειοῦσθαι, εἴπερ ἐν τετταράκοντα ἡμέραις ἢ μικρῷ
πλείοσιν τέλειος. αἰτίαν δὲ ἄν τις ὑπολάβοι τὴν

---

[1] Gaza.   [2] Itali (after Gaza).
[3] Scaliger : τε U.   [4] U : καυλοὶ Heinsius.
[5] U : ταχὺ δὴ Schneider : τάχιστα Wimmer.
[6] U : μόνος Heinsius.   [7] U^{css} : U^{t} omits.
[8] Schneider (νειὸν Wimmer) : νέον U.
[9] Gaza, Itali : ὥρας U.

---

[1] *Cf. HP* 8 9. 3: "There is a difference between lightness
for the soil and lightness for human digestion. For some
are light for the former but heavy for the latter, as legumes
and millet ..."

[2] Legumes are sown in spring.

fruit, their single and shallow root, and their thinly sown seed, all this contributing to their abundant fruit, to their being light for the soil[1] and not exhausting it so much, and to the rapidity of their concoction. For their whole power is directed upward, and there is no counter-pull on the part of the root; this gives us the greater yield and (with the help of the weather)[2] the easy completion of concoction. Indeed where plants have a larger and deeper root and woody fruit they soon exhaust a soil, as chickpea does,[3] which is why it is the only legume that does not renew the soil[4] and needs no ordinary land, but the best.[5]

In the case of this legume its rapid maturing (inasmuch as it matures in forty days[6] or slightly more) looks unreasonable.[7] One would take the

8.4

---

[3] *Cf. HP* 8 9. 1: "Of legumes chickpea most exhausts the soil, although it remains the shortest time in the ground..."

[4] *Cf. HP* 8 7. 2: "Peculiar to chickpea compared to the rest of the legumes is ... its not renewing the soil at all, since it exhausts it ..."

[5] *Cf. HP* 8 7. 2 (of chickpea): "And in general no ordinary soil can support it, but one with black earth and that is fat is needed."

[6] *Cf. HP* 8 2. 6: "... chickpea matures in the smallest number of days, inasmuch as (according to some) it matures in forty days in all from the time of sowing. In any case it evidently matures soonest."

[7] The larger and deeper root should exert a counter-pull and prevent the directing of the whole power upwards and make for slower maturing.

THEOPHRASTUS

ἰσχύν, ᾗ[1] δύναται τῶν τε ἄλλων κρατεῖν καὶ πρὸ
τοῦτο[2] διαρκεῖν, εἴπερ καὶ ὕδατος ἐλαχίστου δεῖ
ται, πρὸς τὸ διαβλαστεῖν[3] μόνον, εἶτα[4] αὐτὸς αὑ
τὸν ἐκτρέφει, εἰ μὴ ἄρα τι καὶ ἡ ἄλμη πρὸς τὴ
τελείωσιν συμβάλλεται, καταξηραίνουσα καὶ ἐξ
ικμάζουσα τὴν ὑγρότητα τὴν πλείω τῆς συμμέ
τρου. φαίνεται δὲ συνεργεῖν πως τῇ γενέσει, κα
οἰκεῖον εἶναι· σημεῖον δ', ὅτι καὶ οἱ πρὸς τῇ θα
λάττῃ βελτίους γίνονται, καὶ ὅτι καταπλυθείσης
τῆς ἅλμης ἀπόλλυται καὶ ἐκζῳοῦται· δι' ἣν δ
αἰτίαν μόνον τούτῳ,[5] σκεπτέον, εἰ μὴ ἄρα τῆς
οὐσίας.

ἀλλὰ δὴ τούτων πέρι[6] τοιαῦταί τινες αἰτίαι.

8.5 μόνα δ' ἀπὸ τῶν ῥιζῶν ἀποφύεται τῷ ὑστέρῳ

[1] Wimmer : εἰ U : ᾗ u : N omits : δι' ἣν aP.
[2] U N aP : τούτω u.   [3] U^c : -σταν U^{ac}.
[4] aP : εἴτε U N.   [5] τούτω U : τοῦτο u.   [6] u : περ U.

[1] The plant's strength would make it attract more food
and lead to the presence of excess fluid.
[2] *Cf. CP* 3 22. 3; 3 24. 3; *HP* 8 6. 5: "Rain is harmless to
pulses except chickpea; for when the brine is washed off
these perish by getting gangrenous and eaten up by caterpillars ..."
[3] Those treated in *CP* 4 8. 1–4.
[4] *Cf. HP* 8 7. 5: "Wheat and barley also grow from the

reason to be its strength, whereby it is able both to prevail over most circumstances and reach maturity by managing without the rest, inasmuch as it also needs very little water, just enough to let it come up, and then supports itself. Unless the brine too contributes something to the maturing by drying out and extracting the fluid that is in excess of the right amount.[1] And the brine does appear to contribute in a way to the production of the plant and to belong to it, the proof being that the plants grown by the sea turn out superior and that when the brine is washed off the plant perishes and engenders worms.[2] (The reason why brine does this in chickpea alone among legumes must be investigated, unless the brine is part of the plant's nature.)

Such are the reasons for these matters.[3]

Grains: A Second Growth[4]

The only grains that come up from the root in the 8.5

roots in many districts in the following year; and they also grow in the same year [reading αὐτοετεῖς with the *CP*; U^c has αὐτο|ετης (ο from ω)] from the plants cut for fodder, another haulm springing up alongside. But the ear of such plants is undeveloped and small. ... They also grow in the following year from plants roughly treated and trodden down so that nothing (so to speak) can be seen, as when an army has passed through. Here too the ears (which they call 'lambs') are small. But no legume can do the like or do it to the same extent."

ἔτει πυρὸς καί κριθή, καὶ αὐτοετεῖς δὲ καὶ ἀπὸ τῶν εἰς κρᾶσιν καρέντων,[1] ἑτέρου καλάμου παραβλαστάνοντος. αἴτιον δ' ὅτι μόνοις ὕπεστι[2] πλῆθος ῥιζῶν καὶ δύναμις, τὰ χεδροπὰ δὲ μονόρριζα καὶ ξυλώδη καὶ ἐπιπολῆς. ὡσαύτως δὲ καὶ ἐπὶ τῶν καταπατουμένων ὑπὸ τῶν στρατοπέδων ὥστε μηδὲν εἶναι δῆλον. ἐξ ἁπάντων δὲ οἱ στάχυες μικροὶ καὶ ἀτελεῖς, ἅτε παλιμβλαστεῖς ὄντες.

9.1 περὶ δὲ τοῦ εἶναι βαρύτερα[3] τὰ χεδροπὰ πρὸς τὰς τροφὰς ἢ τὸν σῖτον ἡμῖν, καίπερ[4] ἐλάττω χρόνον ἐν τῇ γῇ γινόμενα (δοκεῖ δὲ καὶ πρὸς κουφότητα διαφέρειν τοῦτο, διὸ καὶ τῶν πυρῶν οἱ τρίμηνοι κουφότατοι, καὶ τῶν κριθῶν ὁμοίως, ὡς ἐλάττονος ἐνυπάρχοντος τοῦ γεώδους διὰ τὴν βραχύτητα[5] τοῦ χρόνου), τοῖς δὲ ἄλλοις ζῴοις καὶ προσφιλῆ, καὶ ἄλυπα ταῦτ' ἐκείνοις ⟨ἃ⟩ καὶ[6]

---

[1] ego (cf. κειρομένων HP 8 7. 5) : σπαρέντων U.
[2] U^r P : ὕπεστη U^{ar} : ὑπέστη N a.
[3] Schneider (difficiliora Gaza) : βραδύτερα U.
[4] u : καὶ παρ U.
[5] Gaza (brevitatem), Basle ed. of 1541 : βραδυτῆτα U.
[6] ego (⟨δὲ⟩ καὶ Schneider : εἶναι ἃ Wimmer) : καὶ U.

---

[1] Cf. HP 8 2. 3: (all legumes have a single, woody root).

following year are wheat and barley: these also come up in the same year from plants cut for fodder, another haulm growing up alongside the stub. The reason is that these alone have a basis of roots possessing numbers and power, whereas in legumes the root is single, woody[1] and shallow. So too with the wheat and barley trodden down by armies so that no trace of the plants can be seen. But in all these cases the new ears are small and undeveloped, since they are a second growth.

Grains: Comparative Lightness or Heaviness of Cereals and Legumes as Food for Man and Animals: Difficulties

There is the difficulty that legumes are heavier as food for man than cereals, although legumes spend a shorter time in the ground (this spending of a shorter time in the ground being considered to help greatly in also[2] making a plant easy to digest: hence of wheat the three-months variety is the easiest to digest and so too with barley, there being a smaller amount of earthy matter in them because of the short time spent in the ground), whereas animals actually find legumes a pleasure to digest, and food even more indigestible causes them no

9.1

---

[2] As well as in keeping it from exhausting the soil: *cf. HP* 8 9. 1 (of legumes chickpea exhausts the ground most, although it remains in the ground the shortest time).

THEOPHRASTUS

χαλεπώτατα μᾶλλον·[1] καὶ πάλιν ἃ ἡμῖν εὐκατέργαστα, ταῦτα ἐκείνοις χαλεπώτατα, καίπερ ἰσχυροτέροις οὖσιν (πολλὰ γοῦν ἀπόλλυται χορτασθέντα πυρῶν, καίτοι τὸ γεῶδες ἐλάχιστον ἔχουσιν).

9.2 πρὸς δὴ ταύτας τὰς ἀπορίας καὶ εἴ τις ἄλλη παραπλησία ταύταις καθόλου μὲν καὶ κοινὴν τήνδε λαβεῖν χρὴ τὴν ἀρχήν, ὡς οὐ πᾶσι ταὐτὰ[2] πρόσφορα κατὰ τὰς τροφάς, ἀλλ' ἑκάστοις κατὰ τὰς ἰδίας φύσεις. ὅπερ ἐν πολλοῖς μὲν φανερόν, ἐκ πλείστης δ' ἀποφάσεως (ὡς εἰπεῖν) τῶν γε γνωρίμων τὰ[3] περὶ τὰς ἐλάφους αἳ τοὺς ἔχεις ἐσθίουσιν, ὑφ' ὧν τὰ ἄλλα θνήσκουσιν, πολὺ μείζω καὶ ἰσχυρότερα τὴν φύσιν ὄντα· καὶ ἐπ' ἀνθρώπων[4] δὲ τό γε τοιοῦτόν ἐστιν ὥστ' ἀναιρεῖσθαι ταῖς πληγαῖς ὑφ' ὧν οὐθὲν ἄλλοι[5] πάσχουσιν, καθάπερ

---

[1] U : ἡμῖν Wimmer.
[2] ταυτὰ u : ταῦτα U N aP.
[3] [τὰ] Schneider : τὸ Wimmer.
[4] ego : ἀλλών U.
[5] ego : ἄνθρωποι U.

---

[1] Theophrastus apparently has in mind the wheat at Pissangae, for which *cf. HP* 8 4. 5; *CP* 4 9. 5; 4 11. 6.
[2] *Cf.* Nicander, *Theriaca* 139–144; Oppian, *Cyn.* ii. 233,

discomfort, whereas on the other hand food that we digest easily is very indigestible for them, although they are stronger than we (thus many animals die of eating their fill of wheat,[1] which nevertheless has the smallest amount of earthiness).

Solution

To meet these difficulties and the like we must begin at first with a general principle applying to all cases: that the same things taken as food are not wholesome for all animals, but suit the different kinds according to their distinctive natures. This can be seen in many, but the most widely asserted instances (one may say), at least concerning well-known animals, are those of deer eating vipers,[2] which are fatal to other animals much superior to deer in natural size and strength. With man too we have instances of some persons killed by the sting of animals not at all fatal to others, as the persons

9.2

Hal. ii. 209; Aelian, *On the Nature of Animals* ii. 9; *Geoponica* xix. 5. 3: "The deer, by drawing its breath up and in, kills the serpent and draws it to itself"; Lucretius, vi. 765–66; Martial, xii. 29. 5; Pliny, *N. H.* 28. 149: "Everyone knows that deer are fatal to serpents; thus they pull any serpents that are about from their holes and chew them ..." The story rests on a popular etymology of *élaphos* ("deer") from *heleîn* ("capture") and *óphis* ("serpent"): *cf.* Plutarch, *The Cleverness of Animals*, chap. xxiv (976 D); *Etym. Magnum* 326. 2; *Etym. Gud.* 179. 39.

THEOPHRASTUS

[οἱ ὄφεις][1] οἱ[2] ὑπὸ τῶν σκορπίων.

9.3 ὑποκειμένου δ' οὖν τούτου, καὶ τῶν ἄλλων γεω δεστέρᾳ τροφῇ χρωμένων, οὐκ ἄλογον ἐκείνην μᾶλ λον αὐτοῖς ἁρμόττειν, τὴν <δὲ>[3] ἀλλοτριωτέραν εἶναι καὶ ἀπρόσφορον, καὶ διὰ τὴν γλυκύτητα κα τὴν ἡδονὴν πλείω προσενεγκαμένων ῥήγνυσθαι.

τοῦτο μὲν οὖν καθόλου καὶ κοινόν.

ἴδιον δ' ὅτι καὶ συμβαίνει τὰς κοιλίας ἐκπνευ ματοῦσθαι, ἢ πᾶσιν ἢ τοῖς ἔχουσιν ἐχῖνον· φυσῶ δες γὰρ καὶ οὐκ εὐκατέργαστον ὁ πυρὸς ὠμὸς ὤν, ἔτι δ' ἧττον τὸ ἄχυρον. μέγα δὲ καὶ τὸ ἀσύνηθες εἶναι· τὰ γὰρ ξένα, κἂν ᾖ κοῦφα, διαταράττε πλῆθος λαμβάνοντα.

9.4 διὰ δὴ ταύτας τὰς αἰτίας (ὡς ἁπλῶς εἰπεῖν)[4] οὐκ ἀλόγως μὲν ἔχει, δοκεῖ δὲ τῇ παραλλαγῇ.

ὡς δ' ἁπλῶς[5] εἰπεῖν τύπῳ,[6] πρὸς βαρύτητα καὶ κουφότητα τροφῆς καὶ ἡμῖν καὶ τοῖς ἄλλοις

---

[1] ego. [2] U : u erases.
[3] Basle ed. of 1541 : τὴν <δὲ καθαρὰν> Schneider (*purus autem ille* Gaza) : ταύτην δ' Wimmer.
[4] [ὡς ἁπλῶς εἰπεῖν] Schneider.
[5] [ἁπλῶς] Wimmer. [6] <καὶ ἐν> τύπῳ Schneider.

---

[1] *Cf.* Aristotle, frag. 605 (ed. Rose[3]): "Aristotle says in his *Collection of Foreign Observances* that at Latmos in

DE CAUSIS PLANTARUM IV

killed by the scorpions.[1]

At all events when this principle is accepted, and when we add that the brutes use earthier food than ours, the inference is not unreasonable: that food of theirs is the better suited to them; whereas this food of ours is too foreign to their nature and not wholesome, and when they are tempted by its sweetness and the pleasure of eating to consume too much of it, they burst.

9.3

This point, then, is general and of common application.

Of specific application to the case in hand is the following: the digestive tract gets filled with gas either in all the brutes or those that possess the third stomach of ruminants, since wheat taken raw is flatulent and not easy to digest, and the bran still less digestible. The unfamiliarity of the food is also important: strange foods, even when light on the stomach, are upsetting in large quantities.

For these reasons, then, broadly speaking, the difference is not unreasonable, although it is taken to be so owing to the discrepancy in strength of the animals involved.

9.4

Light and Heavy Grains Themselves

To put it broadly in rough outline, the differences in heaviness and lightness both for us and for the

Caria there are scorpions that do no great harm when they sting a stranger, but drive a native to his death."

THEOPHRASTUS

τῆς καὶ φύσεως ἑκάστης οἰκείας,[1] αἱ χῶραι καὶ ὁ ἀὴρ ποιεῖ τὰς διαφοράς· τὰ μὲν γὰρ ἐκ τῶν εὐείλων καὶ εὔπνων καὶ λεπτογείων κουφότατα, τὰ δ' ἐκ τῶν ψυχρῶν καὶ ἐπόμβρων βαρύτερα διὰ τὸ πλείω τροφὴν καὶ ἰσχυροτέραν ἔχειν.

9.5 καὶ διὰ τοῦτο τοῦ μὲν ἄλλου πυροῦ δοκεῖ τοῦ Ἀθήναζε καταπλέοντος ὁ Σικελικὸς ἰσχυρότατος εἶναι, τούτου δὲ ἔτι βαρύτερος[2] μᾶλλον ὁ Βοιώτιος (ὥσπερ ἐν ταῖς ἱστορίαις εἴρηται)· καὶ γὰρ ἡ χώρα πίειρα καὶ ὁ ἀὴρ ψυχρός, καὶ τὰ ἄλλα δὲ τὰ τῶν καρπῶν παραπλησίως ἔχει τῇ βαρύτητι. ἐν δὲ τοῖς Πισσάγγαις[3] ὁ διαρρηγνύων (ὃν εἴπαμεν)

---

[1] ego (*cuiusque naturae* Gaza: τῆς καὶ φύσει ἑκάστων οἰκείας Schneider: τῆς κατὰ φύσιν ἑκάστοις οἰκείας Wimmer): τοῖς καὶ φύσεως ἑκάστης οἰκειους U: τοῖς καὶ φύσεως ἑκάστης οἰκείοις u.

[2] aP: βραδύτερος U N.

[3] ego (U has ταῖς επισυνάγγαις at *CP* 4 11. 6, τοῖς πισσάτοις at *HP* 8 4. 5): τοῖς πιεσάγγαις U.

---

[1] *Cf. HP* 8 4. 5: "... but the Sicilian is heavier than the rest of the wheat imported into Greece, and the Boeotian heavier still. They cite as proof the athletes who can barely consume a choenix and a half in Boeotia, who easily consume two choenixes and a half when they come to

ther animals of the food that *is* suited to the various natures is caused by the country and the air: the produce of sunny and well-ventilated countries with thin soil is the lightest, whereas that of cold and rainy countries is heavier because the grain there has more plentiful and powerful food.

This moreover is why Sicilian wheat is said to be the strongest of all the wheat brought to Athens by sea, whereas the Boeotian is held to be of a still heavier character (as was said in the History),[1] for the land is fat and the climate cold, and the other produce is similarly heavy; and at Pissangae[2] is found the wheat we mentioned[3] that causes burst-

9.5

Athens ... Now the cause of all this lies in the country and the climate."

[2] The name appears as τοὺς Πεσσόγγους (a place or people not far from Pessinus, with which the name appears to be related) in a letter of Eumenes II of 164–163 B.C. (W. Dittenberger, *Orientis Graeci Inscriptiones Selectae* vol. i [Leipzig, 1903], no. 315 A 6). Alexander's army must have traversed the district on the way to Gordium in 334 B.C.

[3] *CP* 4 9. 1; 4 9. 3. *Cf. HP* 8 4. 5 (continuing the citation in note 1): "Indeed in Asia not far from Bactra they say that in a certain district the grain is so big that it reaches the size of an olive pit, and that in the district of the so-called Pissati [read Pissangae] it is so strong that if an animal (reading τι[ς]) took much it burst, and that this happened to a good number of the Macedonians as well." *Cf. CP* 4 11. 6.

THEOPHRASTUS

ὑπερβάλλων δῆλον ὅτι τῇ βαρύτητι. ὁ δὲ Ποντικὸς κουφότατος καὶ σκληρότατος, καίπερ ψυχροῦ τοῦ ἀέρος ὄντος, ὅτι καὶ τὰ σπέρματα καὶ τὰ ἐδάφη κουφότερα, καὶ ἡ χιὼν συνεκπέττει μᾶλλον.

9.6 ὅτι δὲ καὶ τὰ σπέρματα ῥοπὴν οὐ μικρὰν ἔχει φανερὸν ἐκ τῶν τριμήνων καὶ εἴ τινες ἐν ἐλάττονι χρόνῳ τελειοῦνται τούτου· οὐ πολὺ γὰρ τοῦ γεώδους ἕλκοντες, ἀλλὰ κουφοτέρᾳ τροφῇ χρώμενοι καὶ κουφοτέραν ποιοῦσιν καὶ τὴν προσφοράν· ἔοικε δὲ καὶ ἡ ὥρα τι συμβάλλεσθαι,[1] πρὸς τῷ μὴ χρονίζειν, οἷον γὰρ εἰς ὀργῶσαν πίπτειν[3] τὴν γῆν, καὶ παραπλήσιον τὸ συμβαῖνον ὥσπερ τὰ ἐπὶ τὸ ζέον ἐμβαλλόμενα τῶν ἑψομένων· οὐδεμίαν γὰρ οὐδὲ κἀκεῖνα λαμβάνει μώλυσιν.[5] εἰ δέ τινες τῶν ὀλιγοχρονίων[6] τούτων βαρεῖς, ὥσπερ

---

[1] N aP : -άλε- U.
[2] ego : το U.
[3] U : -ει Wimmer.
[4] u : το U.
[5] U : μόλυνσιν u aP : μόλυσιν N.   [6] ὀλιγοχοχρονίων U.

---

[1] *Cf. HP* 8 4. 5: "Now the lightest wheat on the whole is the Pontic ... Odd and in conflict with the lightness of three-months wheat is what happens with Pontic wheats, their hard varieties being spring wheats, their soft, winter wheats, since the soft is much superior in lightness to the hard."

DE CAUSIS PLANTARUM IV

ng, which is evidently heavy to excess. On the other hand the Pontic is the lightest and hardest[1] of wheat, although the climate there is cold, because both the seed[2] and the soil are lighter and the snow helps to bring about better concoction.

That the grains themselves too play no small part is evident from three-months wheat[3] and the varieties that mature even sooner[4]; since they do not attract much earthy substance, but use lighter food, they also make lighter work of consumption. The season[5] too would appear to contribute something, as well as their not tarrying in the ground, since they appear to be sown in the earth when it is (so to say) in heat, and what occurs is like cooking food by immersion in boiling water, for the grains too undergo no half-cooking.[6] But if some of these rapid-growing varieties are heavy, as they say of the

9.6

---

[2] That is, the Pontic variety of wheat.

[3] *Cf. HP* 8 4. 4: "There are many three-months wheats and these are everywhere light."

[4] *Cf. HP* 8 4. 4: "There are also two-months wheats ...; these are ... light to digest and pleasant to eat."

[5] Rapidly maturing wheat was sown in spring, the later (and heavier) varieties were sown at the beginning of winter in early November.

[6] That is, imperfect boiling, which makes them sodden and raw. Such imperfect boiling is a form of lack of concoction: *cf.* Aristotle, *On the Generation of Animals*, iv. 7 (776 a 7–8); *Meteorologica*, iv. 3 (381 a 12).

THEOPHRASTUS

φασὶν τοὺς περὶ Αἶνον,[1] ἐνταῦθα τοῦ σπέρματος τὴν φύσιν αὐτοῦ κατάλοιπον αἰτιᾶσθαι.

καὶ περὶ μὲν τούτων ἀρκεῖ τὰ εἰρημένα.

10.1 τὴν δὲ ἄνθησιν πολυχρονιωτέραν ποιοῦνται τὰ χεδροπὰ τῶν σιτωδῶν, ὅτι τῶν μὲν χνοῶδές τὸ ἄνθος,[2] εἴρηται δὲ ὅτι τὰ ἀσθενέστερα πανταχοῦ θᾶττον τελεοῦνται. διὸ καὶ βρέχεσθαι τὰ μὲν οὐ δύναται, τὰ δὲ καὶ ζητεῖ καὶ ὀνίναται (τάχα δὲ τοῦτό γε[3] κατὰ συμβεβηκός, ὅτι τὸ ὅλον φυτὸν

---

[1] ego (cf. CP 4 11. 4; HP 8 4. 4) : λινον U : λίμνον (λείμνον? u : λιμίον N aP.

[2] τῶν ... ἄνθος ego (flos frumentis imbecillior est Gaza τὸ μὲν σιτῶδες ἀσθενὲς Scaliger : τῶν μὲν ἀσθενέστερον [ἀσθενὲς Wimmer] τὸ ἄνθος Itali) : τῶν μεν σιτῶδες τὸ ἀνθος U.

[3] Scaliger : τε U.

---

[1] Cf. CP 4 11. 4 and HP 8 4. 4: "For such distinctions as these would appear to be the ones most closely connected with the nature of the plants. Here belong the three-months and two-months varieties and any that mature in a still smaller number of days, such as they say is found in the district of Aenos. This gets firm and matures forty days after sowing; and it is said to be strong and heavy and not light like three-months wheat, which is why (they say) it is fed to the slaves; indeed it does not have much bran either."

DE CAUSIS PLANTARUM IV

wheat at Aenos,[1] our recourse here is to find the reason in the nature of the grain itself.

This discussion suffices for these points.[2]

Seed-Crops: Flowering of Legumes and Cereals Compared

Legumes take a longer time flowering than cereals[3] because in cereals the flower is downy,[4] and we have said[5] that whatever is weaker is everywhere sooner matured. This is why the flowers of cereals cannot stand rain, whereas the flowers of legumes not only like rain but are benefited by it.[6] But this benefit, to the flowers at any rate, may be

10.1

---

[2] That is, lightness and heaviness (*CP* 4 9. 1–6).

[3] *Cf. HP* 8 2. 5: "As soon as wheat and barley are set free of the sheath they flower four or five days later and remain in flower for about the same number of days; those that give the highest figure say that they shed the flower in seven days. But the flowering of legumes lasts a long time; the flowering of vetch and chickpea is longer than that of the rest, whereas that of the bean is by far the longest of all, for they say that it flowers for forty days . . ."

[4] *Cf. HP* 8 3. 3.   [5] *CP* 4 1. 4; 2 11. 7.

[6] *Cf. HP* 8 6. 5: "Abundant rain is good for all crops when they have come into leaf and are pregnant with fruit, but injures wheat, barley and the cereals when they are in flower, for the flower is killed. But it is harmless to pulses except chickpea, for chickpea is killed when the brine is washed off, getting gangrenous and eaten by caterpillars . . ."; *cf. CP* 2 2. 2.

εὐτροφεῖ¹ βρεχόμενον, ὁ δὲ καιρὸς² δεῖται βοηθείας πρὸς τὴν τελέωσιν, μάλιστα δ' ὁ κύαμος, ὅτι μανότατον καὶ πολυανθέστατον καὶ πολυκαρπότατον)· ὁ δὲ ἐρέβινθος ἀπόλλυται διὰ τὴν εἰρημένην αἰτίαν.

10.2 ἀλλὰ περὶ μὲν τοῦ σίτου τῆς διαφορᾶς³ λεκτέον. αἰτιάσαιτο δ' ἄν τις μάλιστα τὴν ἀσθένειαν· οὐχ ὑποφέρει γάρ, ἀλλὰ φθείρεται καὶ ἀποπίπτει (καθάπερ καὶ τῶν δένδρων ἐνίων), ὁτὲ μὲν οἷον ἐρυσιβούμενα, ὁτὲ δ' ἐξυγραινόμενα λίαν.

εὐλόγως δὲ καὶ ἡ ἄνθησις οὐχ ἅμα πᾶσι τοῖς μέρεσι· κεχώρισται γὰρ (ἐκτὸς τῶν λοβῶν) καὶ διὰ τὴν εὐτροφίαν· τὰ μὲν γὰρ ἤδη κύει καὶ ἀνθεῖ, τὰ δ' ἄνω προαύξεται⁴ (διὸ καὶ τὰ μὲν αὖα καὶ τέλεια, καίπερ ἐγγυτάτω τῆς τροφῆς ὄντα, τὰ δ'

---

¹ ego : ευτρεφεῖ U : εὐτραφεῖ u.
² U : καρπὸς (*fructus* Gaza) Basle ed. of 1541.
³ U aP (-ὰς N) : διαφθορᾶς u.
⁴ U : προσαύξεται Schneider.

---

¹ *Cf. CP* 2 2. 2.
² *CP* 3 24. 3; *cf. HP* 8 6. 5 (cited in note 6, p. 291).
³ The reason for the inability of the cereal flower to stand rain had been touched on very briefly at *CP* 4 10. 1.

DE CAUSIS PLANTARUM IV

incidental, because the entire plant gets well fed when rained on,[1] and the situation requires replenished resources for the task of completing the formation of the fruit, the bean needing this help most of all, because it has the most open texture and the greatest number of flowers and fruits. Chickpea on the other hand is killed by rain for the reason that was mentioned.[2]

(But we must deal with the difference as it relates to cereals.[3] One would give the weakness of the flowers as the main reason, for they do not bear up against rain but perish and drop (like those of some trees),[4] sometimes getting infected with rust (as it were), and sometimes getting too wet.)

10.2

It is reasonable moreover that there is no simultaneous flowering in all the parts,[5] since the process is rendered progressive (outside of the pods)[6] if only by the good feeding, some parts of the plant being already pregnant with fruit or in flower while the parts above are growing out, which is why some parts are dry or fully developed, although they are closest to the food supply, whereas the parts at the

---

[4] As pomegranate, pear and almond (*CP* 2 9. 3–4).

[5] *Cf. HP* 8 2. 5: "For the flowering of grains with an ear is simultaneous, whereas in plants of the pod-bearing sort and all legumes it is progressive."

[6] Flowering is the first stage of fruiting, and the different seeds in the same pod can therefore be spoken of as simultaneous in a discussion of flowering.

THEOPHRASTUS

10.3 ἐπὶ τῶν ἄκρων χλωρά). (τοῦτο δ' οὐκ ἴδιον, ἀλλὰ καὶ ἑτέρων καὶ πλειόνων κοινόν· πολλὰ γὰρ ἀνθεῖ καὶ γονεύει[1] κατὰ μέρος, καὶ ἅμα τὴν βλάστην ἀφίησιν εἰς τὸ ἄνω· διὸ καὶ τὰ μὲν τέλεα, τὰ δ' ἀτελῆ, τὰ δὲ μέλλοντα.) καὶ οὐκ ἐμποδίζεται τὰ κάτω διὰ τὴν ἐπιρροὴν καὶ τὴν ἀναφορὰν τὴν εἰς τὸ ἄνω, καθάπερ <γὰρ>[2] ὑποκείμενόν τι πᾶσιν ὁ καυλός, ἐξ οὗ τὴν τροφὴν ἔχουσιν ὥσπερ ὀχετοῦ τινος.

ἡ μὲν οὖν ἄνθησις εὐλόγως χρόνιός τε καὶ ἀβλαβὴς αὐτοῖς ὑπὸ τῶν ὑδάτων· ἐν δὲ τοῖς ἄλλοις ὅταν ἀπανθήσῃ σχεδὸν ἐπισινέστερα[3] τοῦ σίτου διὰ τὴν ἀσθένειαν, καὶ μάλισθ' ὁ κύαμος, ἀσθενέστερον γάρ.

καὶ περὶ μὲν τῶν πρὸς ἄλληλα διαφορῶν τοσαῦτα εἰρήσθω.

11.1 περὶ δὲ τῶν ὁμογενῶν, τάχα ἄν τις ἀπορήσειεν

[1] Schneider (*fructificant* Gaza): πονεῖ U.
[2] aP.   [3] Schneider: ἐπισινεστέρα U.

[1] *Cf. HP* 8 2. 5 (continued from note 5, p. 293): "For the first parts to flower are those below; when these have shed their flowers, the parts next to them flower, and in this fashion flowering proceeds toward the top. This is why vetches are pulled up with the fruit already shed in the lower parts while the upper parts are quite green."

extremities are fresh and green.[1] (This is not 10.3
confined to legumes but is found in a good many
other plants as well, many flowering and generating fruit progressively[2] and at the same time
adding to their upward growth, which is why some
fruits are mature, some immature, and some not yet
formed.) And the lower parts are not prevented
from this by the flow of food moving upward, for the
stalk is (as it were) a substrate shared by all the
parts, from which they get their food as if from an
irrigation ditch.[3]

It is reasonable then that the flowering should
last a long time in legumes and be unharmed by the
rains. But in all other matters, once the flower is
shed, legumes are (one might say) even more susceptible to injury than cereals because of their
weakness, the bean especially, since it is weaker
than the rest.

With regard to the differences of cereals and
legumes from one another let this much be said.

*A Difficulty: Different Rates of Growth
Within the Same Kinds*

With regard to seed-crops belonging to the same 11.1

---

[2] *Cf. HP* 7 3. 1 (basil and heliotropion and some wild plants); *HP* 7 14. 2 (anthemon); *cf.* also *CP* 1 11. 7 (citron) and *CP* 5 2. 5.

[3] That is, they are like separate plants fed from the same source: *cf. CP* 1 11. 4.

THEOPHRASTUS

τί ποτ᾽[1] οὐκ ἐν ἴσοις χρόνοις ἅπαντα τελειοῦται, ἀλλ᾽ οἱ μὲν τρίμηνοι τῶν πυρῶν, οἱ δὲ δίμηνοι (καὶ τῶν <κριθῶν>[2] ὡσαύτως)· εἰ δὲ ἐν ἐλάττονί τινες χρόνῳ, πλείων ἡ διαφορὰ πρὸς τοὺς χειμοσπόρους, ὁμοίως δὲ καὶ ἐπὶ τῶν ἄλλων.

μάλιστα δ᾽ ἐπὶ τῶν εἰρημένων (τάχα δὲ καὶ μόνων) αἱ διαφοραί, καὶ τοῦτο εὐλόγως· τὰ μὲν γὰρ ὄσπρια, σπαρέντα χειμῶνος, οὐκ ἂν ὑπομείνειεν διὰ τὴν ἀσθένειαν (πλὴν εἴ τινων ὀλίγων· ὄροβον γὰρ σπείρουσιν καὶ πρώϊον ὥστε μὴ ὥρας ἀπολείπεσθαι[3])· ἡ κριθὴ δὲ καὶ ὁ πυρὸς ἀμφοτέρως ([καὶ][4] ὅσα μὴ τρίμηνα), πλὴν[5] ἐλάττους καὶ ἀσθενεστέρους φέρει τοὺς στάχυς.

τοῦτο μὲν οὖν ὡς καθόλου τῷ γένει πρὸς τὸ γένος.

11.2 τὸ δὲ μὴ ἰσοχρονίους εἶναι κοινὸν καὶ ἐπὶ τῶν δένδρων· ἔστιν γὰρ ἐν ἑκάστοις τὰ μὲν πρώϊα, τὰ δ᾽ ὄψια (καθάπερ ἄμπελος συκῆ μηλέα ἄπιος).

---

[1] U^r N aP : πουτ᾽ U^{ar}.
[2] aP : U N omit (U between lines).
[3] ego (μηδεμιᾷ ὥρᾳ ἐκλείπειν Schneider) : μὴ ὥρα καταλείπεται U.
[4] ego.
[5] μὴ τρίμηνα πλὴν ego (μὲν τρίμηνα τῶν γενῶν Schneider) : μητρην ἀπλὴν U.

kind one might perhaps raise a difficulty: why it is that all do not mature in the same length of time, some wheats maturing in three months, some in two (and similarly with barleys)? And if some wheats mature in still less time [1] the difference with winter-sown wheat is greater, but just as much a difference as in the rest.

The differences occur mainly, perhaps even only, in the seed crops mentioned. [2] This moreover is reasonable: pulse sown in winter would never survive, owing to its weakness (except for some few plants; so farmers sow vetch early [3] as well as late, so as not to run out of it in spring); whereas barley and wheat (except for the three-months varieties) are sown both early and late (still, when sown late [4] they have fewer and weaker ears).

This then is to be taken as a generic difference between the one kind and the other. [5]

But this difference in time is shared with trees, 11.2 for within the several kinds there are some early members, some late, as with the vine, fig, apple and pear. [6]

---

[1] For wheat at Aenos maturing in forty days *cf. HP* 8 4. 4, cited in note 1 on *CP* 4 9. 6.   [2] Wheat and barley.

[3] *Cf. HP* 8 1. 4: "But of legumes such plants as vetch and chickpea are sown at both seasons ..."   [4] *Cf. CP* 4 11. 4.

[5] Of seed crops, namely cereals and legumes; it is not a difference between varieties within the same kind, as of wheats.   [6] *Cf. CP* 1 18. 3.

THEOPHRASTUS

αἴτιον δ' ἐν ἀμφοῖν ὅτι τῆς ἰδίας φύσεως ἡ διαφορά, τὸ δ' ὄνομα κοινόν, ὥσπερ καὶ ἐν τοῖς ζῴοις ἐπὶ τῶν κυνῶν,[1] οὐδὲ γὰρ ἐκεῖνο τὸ γένος ἕν.

ἡ[2] δ' αἰτία παραπλησία καὶ διὰ τί τὰ μὲν πρώϊα, τὰ δ' ὄψια· εἴτε γὰρ θερμότης, εἴτε ψυχρότης, εἴθ' ὅ τι ποτέ, καὶ ἐνταῦθ' ὁμοίως[3] τὸ αἴτιον.

11.3 καὶ καθόλου μὲν οὕτω·

τοῖς δὲ σπέρμασιν καὶ ἐμφανέστερον ἐκ τῶν συμβαινόντων τὸ αἴτιον. ὁ μὲν γὰρ χειμοσπορούμενος

---

[1] Gaza, Basle ed. of 1541 : κοινῶν U.
[2] ἕν. ἡ Wimmer : ενϊ U.
[3] U : ὅμοιον Schneider.

---

[1] *Cf.* Aristotle, *History of Animals*, vi. 20 (574 a 20–29): "The Laconian bitch has a period of gestation of a sixth of a year ... Some bitches have a period of a fifth of a year ... And some of a fourth of a year ..."

[2] *Cf. CP* 1 18. 3–4.

[3] Theophrastus does not commit himself further about the cause in animals and trees. In the late cereals it would appear to be a power of multiplying parts.

[4] Here the distinctive natures of the slow-maturing and fast-maturing varieties: the first has many roots (and haulms), the second few roots (and a single haulm). Theo-

DE CAUSIS PLANTARUM IV

*The Solution: a Difference in Nature
Is Disguised by a Community of Name*

The reason in both cereals and plants is that the difference between early and late belongs to the special nature of the plant, whereas the name is held in common, as in animals is the case with dogs,[1] for here too the kind is not a unit.[2]

The Cause of the Difference

The cause in the animals is like the cause which makes some plants early, some late: whether it is heat or cold or what you will,[3] it is equally present in the plants.

Such then is the general formulation.

11.3

*The Cause Examined in Cereals in
the Light of the Observable Results*

In the seed crops however the cause[4] can actually be seen more clearly when viewed from the results.[5] Thus winter-sown[6] wheat is many-rooted,

phrastus does not push the enquiry further back and ask what causes the many-rooted character, but actually begins at a stage below the "distinctive nature," replacing it by its immediate consequence, the multiplicity of roots.

[5] *Cf. CP* 1 21. 4 (we must consider powers in the light of their results).

[6] Actually autumn-sown, but there is no such compound in Greek.

THEOPHRASTUS

πυρός, πολύρριζος ὤν, καὶ ἐνταῦθα πρῶτον ἀποδοὺς τὴν δύναμιν, ἀποδίδωσι πλῆθος καλάμου (καὶ γὰρ πολυκάλαμος)· ὁ δὲ τρίμηνος καὶ δίμηνος ὀλιγόρριζος καὶ μονοκάλαμος,[1] διὸ[2] τήν τε ἀναφορὰν εὐθὺς ἄνω ποιεῖται καὶ τὴν τελείωσιν ταχεῖαν, ῥᾷον γὰρ τὸ ἔλαττον ἀποτελεῖσθαι.

διὸ καὶ [ὁ][3] πολύχους, ὁ δὲ ὀλιγόχους, καὶ κοῦφος, ὁ δὲ βαρύς, ὥσπερ ἐλέχθη πλείω τὴν τροφὴν ἕλκων καὶ θολωτέραν·[4] ἡ δ' ἀποδρομὴ[5] ταχεῖα κατὰ λόγον ἅτε καὶ μὴ χρονίζοντος ἐν ταῖς ῥίζαις δι' ὀλιγότητα.

11.4 μέγα δὲ καὶ ἡ ὥρα συνεργεῖ· καὶ γὰρ ὁ πολύρριζος, σπαρεὶς ἐν ταύτῃ, θᾶττον ἀναβλαστάνει, καὶ ὀλιγοκαλαμώτερος, καὶ ὁ στάχυς μικρὸς καὶ ὀλιγόπυρος· ὁ δὲ τρίμηνος, πρωϊσπορηθείς, οὐκ ἂν ὑπομείναι[6] διὰ τὴν ἀσθένειαν.

---

[1] ego (ὀλιγοκάλαμος Gaza, Itali : οὐ πολυκάλαμος?) : πολυκάλαμος U.   [2] U : διότι Schneider.
[3] ego : ὁ <μὲν> aP.
[4] U : θολερωτέραν Schneider.
[5] U : ἀναδρομὴ Schneider.
[6] Schneider : ὑπομεῖναι U.

---

[1] Cf. HP 8 4. 4 (3-months wheat is single-haulmed).
[2] The few roots and single haulm.

and after first approving its power by producing many roots goes on to approve it by producing many haulms (since it is a many-haulmed variety too). Three-months and two-months wheat on the other hand is a kind with few roots and a single haulm,[1] which is why it at once directs its growth upwards and rapidly matures: the task with the fewer items[2] is more easily accomplished.

This also is why winter wheat has a big yield,[3] spring wheat a small one and why spring wheat is light on the stomach, winter wheat heavy, attracting as it does a greater amount and muddier type of food[4] (as we said)[5]; and the rapid springing up of the other kind is quite in order, since the roots are too few in number to delay it.

Inhibition by the Season

The season too contributes greatly to the result: thus many-rooted wheat, if sown in the spring season, comes up more rapidly and with fewer haulms, and the ear is small and has few grains; three-months wheat, on the other hand, if sown early,[6] would never survive, it is too weak.

11.4

---

[3] It has many haulms (and ears).
[4] It has more roots; the attraction is therefore stronger and less discriminating.
[5] *CP* 4 9. 4.
[6] In autumn.

THEOPHRASTUS

ἄτοπον δὲ καὶ ὑπεναντίον τὸ συμβαῖνον εἰ βαρεῖς οἱ ὀλιγοχρόνιοι, καθάπερ οἱ περὶ Αἶνόν φασι τοὺς τετταρακονθημέρους·[1] λοιπὸν γὰρ αἰτιᾶσθαι τὴν ἰδίαν φύσιν, ἣν ὁρῶμεν ἐν πλείοσιν μεγάλας ἔχουσαν διαφοράς, καὶ κατὰ τὴν ἔκφυσιν καὶ τελείωσιν, καὶ κατὰ τὰς ἰδίας μορφάς (οἷον μεγέθους καὶ μικρότητος καὶ σχήματος), καὶ τῷ[2] πολύαχυρον εἶναι καὶ μή, καὶ τῷ τέλει[3] δὴ τῷ[4] πρὸς ἡμᾶς καὶ κατὰ τροφὴν καὶ σίτισιν.[5]

11.5 τάχα δὲ καὶ τὰ γένη ποιοῦσιν αἱ χῶραι, ἤτοι πάντα ἢ ἔνια· συνεξομοιοῦσιν γάρ πως ἑαυταῖς ἢ διαφοράν γέ τινα[6] ἐμποιοῦσιν ἢ[7] καὶ πρὸς τὸ δέον χρήσιμος.[8] (ὥσπερ[9] τῷ Θρᾳκίῳ πυρῷ τὸ πολύλοπον εἶναι καὶ ὀψιβλαστῆ· διὰ γὰρ τοὺς χειμῶνας ἄμφω συμβαίνει)· καὶ διὰ ταῦτα καὶ ἐν ταῖς

---

[1] ego (*cuius cura equinoctio delegata est* Gaza : ἰσημερινούς Scaliger) : ἰσημέρους U.
[2] ego : τὸ U.
[3] μή, καὶ τῷ τέλει ego (*incommoditatem* Gaza : μὴ κατωφελῆ Scaliger) : μη κατωτελει U.
[4] δὴ τῶι U : τὸ Wimmer.
[5] Schneider : σίτησιν U.
[6] γε τινὰ U : γενῶν u : γενᾶ N : γένη aP.
[7] ego : ἡ U.

302

DE CAUSIS PLANTARUM IV

An Untoward Result

The result is odd and in conflict with our exposition if rapid-growing wheats are heavy, as they say at Aenos[1] of their forty-day variety, since we must fall back on the distinctive nature and put the heaviness there. We observe that this nature includes important distinctions in a number of matters, distinction in germination and maturing, in the special conformation (as largeness, smallness and shape), in the great or small quantity of the bran, and finally in ultimate utility to man, whether as food or as feed.

The Country Causes the Difference

But perhaps the countries even produce the varieties (either all the varieties or some), since countries bring about a certain assimilation to themselves or at least produce in the wheat a certain difference, useful for meeting the new requirements, just as Thracian wheat is many-coated[2] and a late sprouter, both results being due to the Thracian winters. Hence Thracian wheat, sown early in

11.5

---

[1] *Cf. CP* 4 9. 6; *HP* 8 4. 4.
[2] *Cf. HP* 8 4. 3: "... some wheats have few coats, some many, like the Thracian."

---

⁸ δέον χρήσιμος ego : αεχρησιμος U : χρήσιμος u : ἀχρήσιμος N : ἀχρήσιμον aP.   ⁹ N aP : ὤἐπρ U.

THEOPHRASTUS

ἄλλαις πρωϊσπορούμενος ὁ Θρᾴκιος ὀψὲ δὴ[1] διαβλαστάνει καὶ ἐξαύξεται, καὶ πάλιν[2] ὁ παρὰ τῶν ἄλλων,[3] ἐκεῖ[4] σπειρόμενος [ὀψε],[5] βλαστάνει ·

11.6 γέγονεν γὰρ οἷον φύσις ἤδη τὸ ἔθος. ἐπεὶ καὶ (οἷον ἐν τοῖς[6] Πισσάγγαις[7] καλουμένοις τῆς Ἀσίας) οἱ διαρρηγνύναι[8] λεγόμενοι, καὶ οἱ τὰ μεγέθη τοῖς πυρῆσιν ἴσοι, χώρας ἰδιότητι καὶ φύσει τὰς δυνάμεις ἔχουσι ταύτας, καὶ οὐκ ἂν τηροῖεν[9] μετενεχθέντες. ὃ γὰρ ἐπὶ τῶν ἀμπέλων λέγουσιν, ὡς ὅσα χώρας εἴδη, τοσαῦτα καὶ ἀμπέλων, τοῦτ' ἀληθὲς καθόλου, καὶ οὐχ ἧττον ἐφ' ἑτέρων, ἐὰν ἅμα τῇ χώρᾳ καὶ τὸν ἀέρα τις προσθῇ · διὰ τοῦτο γὰρ αἵ τε ἀνωμαλίαι τῶν καρπῶν,[10] ἀπὸ τῶν αὐτῶν φυτευομένων, αἵ θ' ὅλως ἀκαρπίαι, μὴ φερού-

11.7 σης τῆς χώρας. ἐκ δυοῖν γὰρ ἢ καὶ πλειόνων ὅταν γένηταί τι δύναμιν ἐχόντων, ἀνάγκη κατὰ τὰς

---

[1] ego (Scaliger omits) : δὲ U.
[2] aP : παριν U : πάρι N.
[3] N aP : ἄλλον U.
[4] Scaliger : βεκει U.
[5] ego.
[6] Wimmer : ταῖς U.
[7] ego : επισυνάγγαις U.
[8] u : -ρυ- U.
[9] Scaliger (*nec ... servari incommutabilia possint* Gaza) : οὐ κατηροιεν U.
[10] καρπῶν <τῶν> Schneider.

other countries too, sprouts and heads too late[1]; and again wheat from elsewhere, sown in Thrace, sprouts,[2] ancient habit having by then (so to speak) become nature.[3] Indeed both the wheat reported to cause bursting (as at the place called Pissangae[4] in Asia) and the wheat kernels reported to be as big as olive-pits[5] owe these powers to the special character and nature of the country and would not retain them if introduced elsewhere. For what is said[6] of the vine, that there are as many kinds of vine as there are of country, is true in general, and no less true of other plants if one includes the air with the country. For it is this that accounts not only for the disparities in the crops when countries are planted from the same plants, but also for the failure to produce any (when the country will not bear them).[7] For when something comes about as the result of two or more things[8] possessing power, the whole

11.6

11.7

---

[1] *Cf. HP* 8 8. 1: "Grain taken from countries with severe winters heads too late in places where grain comes up early, and so perishes from drought unless saved by rain."

[2] Instead of remaining dormant, like Thracian wheat.

[3] *Cf. CP* 2 5. 5 with note *a*.

[4] *Cf. HP* 8 4. 5; *CP* 4 9. 5.

[5] *Cf. HP* 8 4. 5.

[6] *Cf. HP* 2 5. 7.

[7] *Cf. HP* 4 4. 1 for an attempt to plant ivy in Babylonia.

[8] The plant is due to (1) the seed and (2) the country (and air).

THEOPHRASTUS

τούτων διαφορὰς καὶ τὸ ὅλον διαφέρειν (ὃ καὶ ἐπὶ τῶν ζῴων συμβαίνει· καὶ γὰρ καὶ τῷ ἄρρενι καὶ τῷ θήλει, καὶ τῇ χώρᾳ καὶ τῷ ἀέρι καὶ ὅλως ταῖς τροφαῖς, λαμβάνουσιν διαφοράς[1])· ὅθεν καὶ γενῶν ἰδιότητες γίνονται, καὶ πολλάκις τὸ παρὰ φύσιν ἐγένετο κατὰ φύσιν, ὅταν χρονισθῇ καὶ λάβῃ πλῆθος.

11.8 ἀλλὰ γὰρ τοῦτο μὲν καθόλου καὶ κοινόν. αἱ δὲ τῶν σπερμάτων διαφοραὶ γίγνονται διὰ τὰς εἰρημένας αἰτίας· ἐπεὶ[2] καὶ τοῦ[3] θᾶττον τελειοῦσθαι παρά τισιν, ὥσπερ ἐν Αἰγύπτῳ φασὶν μηνὶ πρότερον ἢ ἐν τῇ Ἑλλάδι, τὸν ἀέρα τις ἂν αἰτιάσαιτο, μαλακὸν ὄντα καὶ εὐτραφῆ. τὸ δ' ὅλον ἐν δυοῖν τούτοιν θετέον τὰς αἰτίας (ὥσπερ εἴρηται)· ἀέρι καὶ ἐδάφει. τὸ γὰρ αὖ περὶ Μύλας[4] συμβαῖνον

---

[1] U^{cc} from διαφθο- (-ὰς u or U^c).
[2] u : ἐπὶ U.
[3] u aP : τo U : τὰ N.
[4] ego : μῆλον U.

---

[1] For the importance of numbers in deciding a similar question cf. CP 2 17. 3 ad fin.
[2] CP 4 11. 6 ("if one includes the air with the country"); CP 4 11. 7 ("the country and the air").

necessarily varies with the differences in its sources (and this also happens in animals; for animals get differences due not only to the male and the female parent but also to the country and air, in short to their food). From this second source moreover arise peculiarities within kinds, and we often find that what was contrary to nature has become natural, once it has persisted for some time and increased in numbers.[1]

This point, then, is general and of common application. But the differences within the kinds of grains are brought about by the causes mentioned[2]; so one would give the air as the cause for earlier ripening in some places, as ripening is said to be a month earlier in Egypt than in Greece,[3] the air in Egypt being mild and nutritious.[4] Indeed in general we must lay it down that the place to look for the causes[5] is these two things (as we said)[6]: air and soil. So in the case again of what results at

11.8

---

[3] *Cf. HP* 8 2. 7: "Country differs from country and air from air in effect on maturation as well. For some countries are held to produce in less time, as Egypt most markedly among the rest: barley there is harvested in six months, wheat in seven, whereas in Greece barley is harvested in the seventh month (but in most districts in the eighth), wheat still later."

[4] It is foggy and laden with dew: *CP* 6 18. 3; *HP* 8 6. 6.

[5] That is, the causes proceeding from the country (*CP* 4 11. 6).

[6] In the passages cited in note 2.

THEOPHRASTUS

τῆς Σικελίας,[1] ὥστε τὸν ὕστατον σπείροντα θερίζειν ἅμα τοῖς πρώτοις, ἐπὶ τὴν χώραν ἀνοιστέον ὡς εὔτροφον, ὁ γὰρ ἀὴρ παραπλήσιος.

11.9 τὸ γὰρ μὴ ἰσοχρονεῖν τὰ σπέρματα, καθάπερ καὶ τὰ ζῷα, παρὰ πᾶσιν, ὁποιασοῦν οὔσης τῆς ὥρας,[2] οὐδὲν ἄτοπον. ἐκεῖνα μὲν γὰρ ἐν ἑαυτοῖς ἔχει τὰς ἀρχὰς τὰς κυριωτάτας, τὸ δὲ σπέρμα καὶ ὅλως ἐν τῷ ἀέρι μᾶλλον, εἰ δὲ μή, τάς γε πρὸς βλάστησιν, καὶ ὅλως γένεσιν καὶ διαφθοράν.[3] διόπερ οἷον ἂν ᾖ τὸ ἔτος, ἀκολουθεῖ καὶ τὰ τῶν καρπῶν, ἔν τε τοῖς ἄλλοις, καὶ ἐν τῇ πρωϊότητι καὶ ὀψιότητι.

11.10 σύμφωνον δὲ τρόπον τινὰ καὶ οὐ πόρρω τούτων καὶ τὸ μὴ κινεῖσθαι πρότερον μήτε σπέρμα μηδὲν μήτε φυτόν, ἀλλὰ κατὰ τὴν οἰκείαν ὥραν.[4] ὃ καὶ θαυμάζεται περὶ τῶν σπερμάτων, ὅτι διαμένει

---

[1] ego : τελέσεως U.
[2] ego : χώρας U.
[3] Gaza : διαφοράν U.
[4] Heinsius : χώραν U.

---

[1] *Cf. HP* 8 2. 8: "It is ... said that in Sicily at the place called Mylae ... the crops sown late mature pretty fast: pulses are sown for six months, but the farmer who sowed in the last month harvests his crop with those who sowed

Mylae in Sicily,[1] where the last sower harvests his crop with those who sowed first, we must trace the result to the land and its fertility, since the air is much the same as elsewhere in those parts.

For it is not at all odd that grains do not behave like the animals and take the same time in all places for their gestation, whatever the character of the season when the seed is sown: animals have their most important starting-points in themselves, whereas the seed of the plant gets its impulses in general from the air (at least the ones that make it sprout and in general bring it to birth or destroy it). This is why "as goes the year, so goes the corn"[2] both in other matters and in that of coming up early or late.

11.9

A Related Problem: Why Seeds Survive Until Their Time

In agreement with this in a way and not far removed from it is this: that no seed and no plant begins to stir before its proper season.[3] This has aroused surprise in the case of the seeds, because

11.10

---

first ..." (At Mylae one can sow five months later than the first to sow elsewhere in Sicily, and harvest the crop at the same time as they.)

[2] A paraphrase of the proverb "The harvest is the year's and not the field's" (*CP* 3 23. 4; *HP* 8 7. 6).

[3] *Cf. CP* 1 10. 6; *HP* 7 1. 7; *HP* 7 10. 1.

πρὸς τὸ θέρος ἔνια καὶ οὐ διαφθείρεται, πολλῶν
ὑδάτων καὶ εὐδιῶν γενομένων· ἐπεὶ τό γε μὴ
βλαστάνειν ἧττον ἄλογον, μὴ ἔχοντα τὴν οἰκείαν
κρᾶσιν. αἴτιον δὲ ταὐτό[1] πως ὑποληπτέον, ἐς
ἀσφαλὲς τῆς φύσεως πρὸς ἄμφω τιθεμένης ἐν τῇ
τῶν περιεχόντων ἰσχύϊ· φαίνεται γὰρ τὰ μὲν ξυλώδη (καθάπερ τὰ δενδρικά[2]), τὰ δὲ πολυχίτωνα,
τὰ δ' ἄλλας τοιαύτας ἔχοντα φυλακάς.

καὶ ταῦτα μὲν δὴ κοινά πως τῆς φύσεως.

12.1   ὑπὲρ δὲ τῶν σπερμάτων, πῶς ποτε τὰ τεράμονα καὶ ἀτεράμονα γίνεται; πότερα διὰ τὴν χώραν, ἢ διά τιν' ἀέρος κατάστασιν,[3] ἢ δι' ἄλλο τι
πάθος, καὶ πάντα ἢ ἔνια (δοκεῖ γὰρ δὴ καὶ μάλιστα ἐπὶ τῶν κυάμων καὶ φακῶν); συμβαίνει δὲ
πολλάκις καὶ τὸ χωρίον ὁτὲ μὲν τεράμονα φέρειν,
ὁτὲ δὲ ἀτεράμονα, τῆς αὐτῆς ἐργασίας τυγχάνον·

---

[1] ego (τούτου Schneider): τοῦτο U.
[2] Wimmer (*arborum* in his translation): ἄνθηκα U.
[3] ego (*cf. HP* 8 8. 7 ἀέρος κατάστασίς τις; διὰ τὸν ἀέρα καὶ τὰς <ἰδίας κράσεις καὶ διαθέσεις> Heinsius): δια τον αέρα και τὰς U.

---

[1] For examples of seeds that do not sprout until summer or later *cf. HP* 6 2. 6; 6 4. 4; 6 5. 1; 6 5. 2; 6 5. 4; 7 10. 1.

some of them survive until summer[1] and do not perish, in spite of the many rains and spells of fine weather in the interval; as for their not sprouting, this is less unreasonable, since they do not have the properly tempered weather. But we must take the cause to be in a way the same: the nature of the plant stores the seed away against both eventualities[2] in the strength of the enclosures. For we see that some seeds are woody (as in trees), some have many coats, and others have similar protections.

These matters, then, depend in a way on the nature of the plant as well.

A Problem: Ready and Stubborn Seeds

Touching seed-crops, how do the ready and stubborn[3] ones arise? Is the difference due to the country or to a certain settled state of the air or else to something in their character? And is it present in all kinds of seed crops or only in some (for it is believed that it occurs especially in bean and lentil)[4]? It often happens that a field bears ready seeds at one time, stubborn ones at another, although the

12.1

---

[2] Destruction and sprouting.

[3] A ready (or easy or yielding or biddable or responsive or cooperative) seed is one readily softened by boiling (*CP* 4 12. 2); the stubborn (or intractable or obdurate or uncooperative) one tends to remain hard. *Cf. HP* 8 8. 6–7.

[4] *Cf. HP* 8 8. 6, cited on *CP* 4 12. 13.

THEOPHRASTUS

καὶ τῶν συνεχῶν, αὔλακος[1] μόνον διειργούσης,[2] τὸ μὲν ἀτεράμονα, τὸ δὲ τεράμονα· καὶ τῶν σπερμάτων ὁτὲ μὲν ἐκ τῶν ἀτεραμόνων τεράμονα[3] γίνεσθαι, ὁτὲ δὲ ἐκ τῶν τεραμόνων ἀτεράμονα· κατὰ δὲ τὴν ἔκφυσιν καὶ βλάστησιν, καὶ τὴν ἁδρότητα καὶ εὐκαρπίαν, οὐδὲν διαφέρει τὸ ἀτέραμον ὡσάν τι νενοσηκὸς ἢ πεπονηκός.

12.2 ὑπὲρ δὴ τούτων, καὶ εἴ τι ἄλλο συνάπτει πρὸς τὴν ἀπορίαν ταύτην, πρῶτον ἐκεῖνο λεκτέον· ὅτι τὸ τέραμον καὶ ἀτέραμον πρὸς τὴν πύρωσιν λέγεται καὶ διάχυσιν, καὶ (ὡς ἁπλῶς εἰπεῖν) πρὸς τὴν τροφὴν τὴν ἡμετέραν. τὸ μὲν γὰρ εὐδιάχυτον καὶ τῇ ἑψήσει ταχὺ ἀλλοιούμενον, τέραμον· τὸ δ' ἀδιάχυτον ἢ ἀναλλοίωτον ἢ βραδέως ἀλλοιούμενον, ἀτέραμον. τοιοῦτον δὲ ἑκάτερον, εἰ τὸ μὲν μανὸν εἴη καὶ μαλακόν, τὸ δέ, πυκνὸν καὶ σκλη-

---

[1] U^r N aP : -ας U^ar.
[2] u : δ' εἰργούσης U.
[3] U^r N aP : ατεράμονα U^ar.

[1] *Cf. HP* 8 8. 7: "Again a certain settled state of the air causes this sort of difference (*i.e.*, that between ready and stubborn seeds); proof of this is that the same farms, worked in the same way, now produce ready seeds, now

DE CAUSIS PLANTARUM IV

farming has been the same[1]; and that of adjoining fields, separated by the breadth of a furrow, the one bears stubborn seeds, the other ready[2]; and, turning to the seeds, that the stubborn seeds sometimes produce ready ones, the ready sometimes stubborn ones. In emergence from the seed and sprouting, and in the stoutness and productivity of the plant, nothing distinguishes the stubborn sort as having suffered from any disease or hardship.

A Preliminary to Solution: The Meaning of the Terms

About these matters and any others that touch on this difficulty we must begin by saying that the distinction of ready and stubborn is made with reference to the exposure of the seed to fire and the loosening of its structure, and broadly speaking with reference to human consumption, the seed that is easily loosened and quickly altered by boiling being ready, the seed not loosened or altered or altered only slowly being stubborn. Each will have this character under the following conditions: the one if it is open in texture and soft, the other if it is

12.2

stubborn."
[2] *Cf. HP* 8 8. 7: "... and of farms some lying next to each other and similarly situated and with no difference in soil bear the one ready seeds, the other stubborn, sometimes with a furrow's breadth between them."

THEOPHRASTUS

ρόν· οὕτως ἂν τὸ μὲν δέχοιτο τὴν θερμότητα καὶ ὑγρότητα δι' ὧν ἡ διάχυσις, τὸ δ' οὐ δέχοιτο, ἀλλ' ἀποστέγοι τῇ πυκνότητι καὶ σκληρότητι.

12.3   διαιρουμένου δ' οὕτω τοῦ τεράμονος καὶ ἀτεράμονος, σκεπτέον παρὰ τίνας καὶ ποίας τινὰς αἰτίας ταῦτα συμβαίνει.

τὸ μὲν οὖν ἁπλοῦν ἐκεῖνο, καὶ ἀληθές (ὃ καὶ ἐπὶ τῶν πρότερον εἴρηται) · διότι παρὰ τὴν τροφὴν τὰ τοιαῦτα γίνεται πάντα, τῷ ποιάν[1] τε καὶ ποσὴν[2] εἶναι· μεθίστησι γὰρ αὐτή.[3] συμβαίνει γὰρ τὰ μὲν ἐν τοῖς ἀλεεινοῖς καὶ διακόποις καὶ λεπτογείοις καὶ ἡλιοβόλοις κούφην τε τὴν τροφὴν καὶ εὐκατέργαστον ἔχειν, ὥστε καὶ τὰ ξυνιστάμενα μανὰ καὶ μαλακὰ γίνεσθαι.

διὰ τοῦτο γὰρ καὶ τὰ Λήμνια[4] τεράμονα, διότι τοιαῦτα τὰ ἐδάφη (τὸ γὰρ ὅλον τὴν γῆν τεράμονά τινες καλοῦσιν τὴν τοιαύτην, ἐν δὲ τῇ τεράμονί φασι γίνεσθαι ‹τεράμονα›[5]) · τεράμονα δὲ καὶ τὰ

[1] u : ποίαν U.   [2] u : πόσην U.   [3] aP : αὐτὴν U^{ac} (-ήν U^c) : αὐτή u : αὐτῆ N.   [4] U^{cc} : λι- U^{ac}.
[5] added by Schneider (following Gaza) after τεράμονι : placed here by me.

---

[1] The causes are different characters in ground, air and seed.

DE CAUSIS PLANTARUM IV

close in texture and hard; for the former will then admit the heat and fluid that bring about the loosening, and the latter will not admit them but keeps them out by its close texture and hardness.

The "ready" and the "stubborn" being distinguished in this way, we must consider the nature and character [1] of the causes that bring them about.

12.3

(1) *Causes in the Types of Ground*

Now that bald statement (which we also applied [2] to earlier problems) is not only bald but true, that all such differences [3] arise from difference in quality and quantity of food, since it is the food that changes the character. For it happens that in warm, well-manured, light and sunny ground the plants get light and easily processed food, so that the seeds that are formed are open in texture and soft.

So the reason why Lemnian seeds [4] are ready is that the soil of Lemnos is of this character [5] (some persons even go on to call such ground "ready," asserting that seeds "come out ready in ready ground"). The seeds are ready in Egypt too, both

---

[2] *CP* 4 9. 4 (explaining lightness and heaviness), *CP* 3 17. 7 (explaining the change from acid to sweet).

[3] Difference of character in the same plant, not differences that make another plant.

[4] Not mentioned elsewhere.

[5] The ground of Lemnos is covered with volcanic ash.

315

THEOPHRASTUS

ἐν Αἰγύπτῳ διά τε τὸ ἔδαφος καὶ διὰ τὸν ἀέρα, τὸ γὰρ θερμὸν οἰκεῖον τῇ τεραμότητι (καὶ ὅλως τῇ πέψει · διὸ καὶ τὰ κοπριζόμενα προτερεῖν φασιν τῶν ἀκοπρίστων σχεδὸν εἴκοσιν ἡμέραις).

12.4 τὰ μὲν οὖν ἐν τῇ ἀλεεινῇ καὶ κούφῃ διὰ τὰς αὐτὰς αἰτίας τεράμονα · τὰ δ' ἐν τῇ ψυχρᾷ καὶ πιείρᾳ¹ καὶ γλίσχρᾳ καὶ ὥσπερ κεραμέᾳ,² καὶ ἔτι δὴ τῇ λειμωνίᾳ καὶ ἐφύδρῳ καὶ ἑλώδει, πάντα [τεράμονα]³ διά τε τὸ πλῆθος καὶ τὴν ἰσχὺν τῆς τροφῆς πυκνά τε⁴ καὶ βαρέα καὶ σκληρά, τοῦ γεώδους τε πολλοῦ καὶ τοῦ ψυχροῦ καταμιγνυμένων (ἡ γὰρ δὴ πῆξις καὶ ἡ πύκνωσις ἐκ τούτων, ἐξ ὧνπερ καὶ ἡ σκληρότης) · ἐν δὲ τοῖς ἑλώδεσιν καὶ ἐφύδροις ἀχρεῖα τὸ ὅλον · οὐ γὰρ ἐνδιδοῖ βρεχόμενα,

---

¹ Coray (*spissa* Gaza) : πικρὰ U.
² ego : κεραμία U.
³ ego (*Omnia haec incoctilia* Gaza) : πάντα (πάντ' Schneider) ἀτεράμονα Scaliger.
⁴ U : δὲ Schneider.

---

¹ That is, the warm climate.
² Manure is hot: *cf. CP* 3 6. 1.
³ That make them open in texture and soft (*CP* 4 12. 3); except for the seeds of Lemnos and Egypt they have not yet been actually termed "ready."
⁴ The stiffening comes from the cold, the closing of texture from the earthy.

because of the ground and because of the air,[1] since heat is favourable to readiness (and to concoction in general, which is why plants receiving manure are asserted to mature some twenty days ahead of those receiving none).[2]

The seeds, then, in warm and light ground are for the same reasons[3] ready; whereas the seeds in cold and rich and viscous ground, that consists (as it were) of potter's clay, and again in meadow land and land with surface water and swampy, are all of them, owing to the great quantity and strong quality of their food, close in texture, heavy and hard, since a great deal of the earthy and cold has entered into their composition (for the stiffening and closing of texture come from these,[4] and from these in turn comes their hardness), and in swampy land and land with surface water are simply useless,[5] since they do not give when immersed in water,[6] and this

12.4

---

[5] Useless that is for cooking (and human food), and hence stubborn; the ones just mentioned have only been called close in texture, heavy and hard, and this may not have excluded the possibility of cooking some of them.

[6] That is, the amount and strength of the fluid in them is so great that it is not affected by the water in which they are boiled. *Cf.* Aristotle, *Meteorologica*, iv. 3 (380 b 20–21) (the fluid in things boiled is brought out of them by the heat in the cooking water); iv. 3 (381 a 4–8) (to be capable of being boiled things must be capable of being thickened or becoming smaller in amount or heavier).

THEOPHRASTUS

διὸ καὶ καταχρῶνται πρὸς τὰς ὗς.

αἱ μὲν οὖν ἐκ τῶν ἐδαφῶν αἰτίαι σχεδὸν αὗται.[1]

12.5 ἐπεὶ δὲ καὶ τὰ ὕδατα τὰ οὐράνια καὶ ὁ ἀὴρ συνεργεῖ ταῖς τροφαῖς, διὰ τοῦτο ταῖς τε[2] ἐπομβρίαις ἀτεράμονα μᾶλλον γίνεται, πλέονος οὔσης καὶ ἀπεπτοτέρας τῆς τροφῆς (σχεδὸν γὰρ ταὐτὸ[3] συμβαίνει καὶ εἰ[4] ἐκ τῆς τοιαύτης γῆς), καὶ εἰ τοιαύτη τοῦ ἀέρος ἡ ψυχρότης ὥστε ἔνυδρός τις εἶναι καὶ μὴ πνευματική· πῆξιν γὰρ οὕτω ποιήσει καὶ πύκνωσιν.

διῃρημένων[5] δὲ εἰς ταῦτα τῶν αἰτίων,[6] οὐδὲν κωλύει καὶ τὸ αὐτὸ χωρίον ὁτὲ μὲν τεράμονα φέρειν, ὁτὲ δὲ ἀτεράμονα, μὴ ὁμοίων[7] ἀλλ' ἐναντίων γινομένων τῶν ἐκ Διός, καὶ τῶν διειργομένων αὔλακι τὰ μὲν τεράμονα, τὰ δὲ ἀτεράμονα γίνεσθαι,

§ 5 lines 12–13 Plutarch, Quaest. Conv., vii. 2. 3 (701 C-D): οὐ δεῖ δὲ θαυμάζειν ἀκούοντας τῶν γεωργῶν ὅτι καὶ δυεῖν αὐλάκων ἡ μὲν ἀτεράμονας <ἡ δὲ τεράμονας> ἐκφέρει τοὺς καρπούς.

[1] u : αὐταί U.   [2] U N : γε aP.
[3] ego : τοῦτο U.   [4] u : ἡ U.
[5] u aP : διειρημένων U N.
[6] u : αἰτίων U : αἰτιῶν N aP.
[7] u : ὁμοίωι U.

DE CAUSIS PLANTARUM IV

is why they are fed to the pigs.

These, then, are the causes (one might say) that come from the types of ground.

(2) *Causes in the Types of Air*

Since rain and the air also cooperate in the type of feeding, the seeds tend more to become stubborn in rainy spells, the food being then more copious and so more unconcocted (for the same result, one might say, occurs then as when the food comes from ground with surface water)[1]; again the seeds are more stubborn when the coldness of the air is of a damp sort and not windy, since under these conditions the air will produce stiffening[2] and closing of texture.

12.5

The Problem Solved by the Two Kinds of Causes

Now that the causes have been distinguished into these two[3] there is nothing to prevent (1) the same field from bearing ready seeds at one time and stubborn at another,[4] when the weather conditions are not similar but the opposite, and (2) crops separated by a furrow's breadth from becoming the one ready, the other stubborn,[4] when the ground is

---

[1] *CP* 4 12. 4.
[2] *Cf. CP* 5 12. 7 and *CP* 4 12. 4 with note 4.
[3] From the ground and from the air.   [4] *CP* 4 12. 1.

εἴπερ τοῦ μὲν τοιόνδε, τοῦ δὲ τοιόνδε τὸ ἔδαφος.

12.6 ὥσπερ γὰρ ἐν τοῖς μετάλλοις ῥάβδους, οὕτως κἂν[1] τοῖς ἐργασίμοις ὑπολαβεῖν χρὴ διατετάσθαι τὴν μὲν τοιάνδε, τὴν δὲ τοιάνδε, παρ' ἀλλήλας, ὥστε, τῶν αὐτῶν ἐκ τοῦ ἀέρος γινομένων, ταῖς ἰδίαις δυνάμεσι ποιεῖν τὰς διαφοράς, ὁτὲ δὲ καὶ τοῦ αὐτοῦ χωρίου μικρόν τι μέρος εἶναι τοιοῦτον· ὡσαύτως δὲ καὶ εἴ τις ἀέρος ἢ πνεύματος προσπέσοι τοιαύτη ψυχρότης ὥστε ποιῆσαι πῆξιν ἢ καθ' ὅλον ἢ κατὰ μόριον (ὃ καὶ κατὰ τὴν φθορὰν συμβαίνει, τῷ τὰ μὲν παραλλάττειν, τὰ δὲ ἐπιβαίνειν).

καὶ εἴ τις ἐν αὐτοῖς τοῖς σπέρμασιν ἀνομοία διάθεσις ὑπάρχει· τὰ γὰρ ἀσθενέστερα δῆλον ὡς εὐπαθέστερα, τὸ δ' ἀσθενὲς κἂν φύσει γίνοιτο, τῷ λαμβάνειν τινὰ μεταβολὴν ἐν ἑαυτῷ.

12.7 διὸ καὶ τῶν ἐπὶ τοῦ καυλοῦ κυάμων οὐδὲν κωλύει τὸν[2] μὲν[3] ἀτεράμονα τῶν λοβῶν[4] εἶναι, καὶ

§ 7 lines 1–4 *Cf.* Plutarch, *Quaest. Conv.*, vii 2. 3 (701 D) (continued from *CP* 4 12. 8): καὶ ὃ μέγιστόν ἐστι, τοῦ κυάμου (τοὺς κυάμους Wyttenbach) τῶν λοβῶν οἱ μὲν τοίους, οἱ δὲ τοίους, δηλονότι τοῖς μὲν ἧττον τοῖς δὲ μᾶλλον ἢ πνεύματος ψυχροῦ <προσ>πεσόντος <ἢ> ὕδατος.

DE CAUSIS PLANTARUM IV

of the one character in the one field, of the other in the other. For just as with veins in a mine, so in cultivated land we must suppose that different kinds of soil extend alongside one another, so that although the weather conditions are the same, the different soils, by the powers that distinguish them, produce the difference in the seeds; and that sometimes even in the same field a small portion belongs to such a different vein. So too if coldness of air or wind should strike of a sort to produce stiffening in the whole field (or in a portion, this also happening when the cold kills, because the wind by-passes one spot and overflows another).[1]

(3) *A Cause in the Seed Itself*

Again there might be a dissimilar disposition in the seeds themselves. For the weaker ones are evidently the more easily affected, and weakness could also arise naturally, by the seed's undergoing a certain change within itself.[2]

This is why there is nothing to prevent, among beans on the same stalk, the one podful from being

---

[1] *Cf. CP* 5 12. 10.
[2] And not attributable to the soil or air.

---

[1] aP : καὶ U N.
[2] ego : τῶν U.
[3] μὲν <τεράμονα, τῶν δὲ> Itali (after Gaza).
[4] U : τὸν λοβὸν u.

THEOPHRASTUS

ἐν τῷ αὐτῷ λοβῷ (καθάπερ τινές φασιν, εἴπερ λέγουσιν ἀληθῆ)· τὸν μὲν γὰρ ἀσθενέστερον εἶναι, τὸν δὲ ἰσχυρότερον ἐνδέχεται (τῆς[1] ἀσθενείας τοῦ κυάμου κἀκεῖνο σημεῖον ἄν τις λάβοι· μόνος γὰρ δοκεῖ μεταβάλλειν τὴν χρόαν ἐκ λευκοῦ μέλας).

12.8 ὅτι δὲ πῆξίς τις καὶ πύκνωσίς ἐστιν ὑπὸ τοῦ ψυχροῦ δι᾽ ἣν ἀτεράμονα γίνεται, μαρτυρεῖ καὶ τὸ περὶ Φιλίππους συμβαῖνον περὶ τοὺς κυάμους· πνεῖ[2] γὰρ σφόδρα ψυχρὸν[3] καὶ λικμωμένοις ἐπιγινόμενον·[4] ἐὰν δ᾽ ἐν[5] τοῖς ἀχύροις ἀπηλοημένοις[6] ἐν τῇ ἅλῳ καὶ καθαροῖς οὖσιν, οὐ μεταβάλλουσιν, ἀλλὰ τεράμονες,[7] ὁτὲ μὲν γὰρ ὑπὸ τῶν ἀχύρων, ὁτὲ δὲ ὑπὸ τῆς πρὸς ἀλλήλους συναφῆς σκεπάζονται, καὶ ἅμα τῆς γῆς θερμότητος· ὅταν

---

§ 8 lines 2–8 Plutarch, *Quaest. Conv.*, vii 2. 3 (701 C): ἐνιαχοῦ δὲ καὶ πνεῦμα λικμωμένοις ἐπιγινόμενον ἀτεράμονας ποιεῖ διὰ τὸ ψῦχος, ὥσπερ ἐν Φιλίπποις τῆς Μακεδονίας ἱστοροῦσι· τοῖς δ᾽ ἀποκειμένοις βοηθεῖ τὸ ἄχυρον.

lines 2–15 Pliny, *N. H.* 18. 155: circa Philippos ateramum nominant in pingui solo herbam, qua faba necatur, teramum, qua in macro, cum udam quidam ventus adflavit.

[1] τῆς <δ᾽> aP.
[2] ego (ἐκεῖ Heinsius): ἐστι U.
[3] ψυχρὸν <ἐκεῖ τὸ πνεῦμα> Itali (after Gaza).

stubborn, and so with the beans in the same pod, as some assert (supposing the assertion true); for it is possible for the one bean or podful to be weaker, the other stronger. As for the weakness of the bean, one could take the following as further proof: that the bean alone, it is believed, changes its colour from light to dark.

Other Cases of Readiness and Stubbornness Explained by These Causes

But that what makes the seeds stubborn is a stiffening and closing of texture brought about by cold is attested by what occurs to the beans at Philippi. A very cold wind rises there and blows on the beans during the winnowing too.[1] If it blows on them as they lie threshed on the threshing floor among the chaff and stripped of the pod, they do not change but remain ready, for they are sheltered partly by the chaff and partly by their contact with one another, and then too by the warmth of the ground. Whereas

12.8

---

[1] *Cf. HP* 8 8. 7: "At Philippi if the bean when it is being winnowed is caught by a local wind, ready beans become stubborn."

---

[4] λικμωμένοις ἐπιγινόμενον Plutarch : ἀτεράμονες τι γίνονται U.

[5] U^c : ἐ U^{ac}.

[6] Schneider : ἀπηθαλωμένοις U : ἀποκειμένοις Plutarch.

[7] P : ἀλλἀτεράμονες U : ἀλλ' ἀτεράμονες u N a.

THEOPHRASTUS

δὲ μετέωροι ληφθῶσι, τό τε πνεῦμα μᾶλλον ἰσχύει καί, οὐδαμόθεν ἐχόντων σκέπην, εἰσδύεται καὶ πήγνυσιν· ἅμα δὲ καὶ ἀσθενέστατοι τότε γίγνονται, γυμνούμενοι πρῶτον τῶν ἀχύρων καὶ τῆς θερμότητος[1] τῆς περιεχούσης· αἱ δὲ μεγάλαι μεταβολαὶ μάλιστα κινοῦσιν.

12.9 ἐπιγενομένου δὲ χρόνου (καὶ ὥσπερ ἤδη συνεστηκότων) ἐὰν δικμῶνται, πάλιν μηδὲν πάσχειν ὑπὸ τοῦ πνεύματος· ἰσχυρότερον [τε][2] γὰρ ἤδη καὶ τὸ κέλυφος τὸ περιέχον.

τὸ δὲ πλείω χρόνον ἐᾶν[3] ἡλοημένους ὑπαιθρίους, ποιεῖν[4] ἀτεράμονας (ὥσπερ τινές φασιν) οὐκ ἄλογον· ἀποψύχονται γὰρ δῆλον ὅτι μᾶλλον τοῦ συμμέτρου καὶ ἅμα κατασκληρύνονται.

ἀκόλουθον δέ πως τούτῳ καὶ ὅτι τὸ μετὰ τὴν σπορὰν εὐθὺς ἐπιγινόμενον ὕδωρ ἀτεράμονας ποιεῖ· ἀσθενῆ γὰρ αὐτὸν λαβὸν[5] ἐν τῷ διαβλαστάνειν ὄντα, κατέψυξεν· ὃ πρὸς μὲν τὴν φύσιν[6]

---

§ 9 lines 5–6 *Cf.* Plutarch, *Quaest. Conv.*, vii 2. 3 (701 C): τοὺς δὲ καρπούς, κἂν ἐπὶ τῆς ἅλω διαμείνωσι πλείω χρόνον ὑπαίθριοι καὶ γυμνοί, μᾶλλον ἀτεράμονας γίνεσθαι λέγουσιν τῶν εὐθὺς αἱρομένων.

[1] u : θερτητος U. [2] Schneider.
[3] u aP : ἐὰν U : ἐὰν N. [4] u : ποιεῖ U.
[5] u : λαβὼν U aP : λοβὼν N.

if the wind catches them off the ground, the wind has greater strength and enters and stiffens them when they have no shelter on any side; and at the same time the seeds then reach their weakest state, since they are now for the first time divested of the chaff and of the warmth that had hitherto enveloped them, and it is the great changes that have the greatest effect.[1]

But if the beans are winnowed after an interval and have had time to become (as it were) firm after the threshing, the wind now does them no harm, since even the skin round them has got stronger.

But that leaving the threshed beans lie in the open much longer makes them stubborn (as some assert) is not unreasonable, since they evidently cool off more than is good for them and at the same time get too hard.[2]

It accords in a way with this that rain right after the sowing makes them stubborn, for it catches the seed at a weak moment, engaged in sprouting, and chills it; and although the occurrence does no harm

---

[1] One may suspect that the labourers preferred to await a warmer wind for the task of winnowing and devised this excuse for the delay.

[2] The cold wind carries the fluid away with the heat: *cf. CP* 5 12. 6. The hardening is observable; the cooling is inferred.

---

[6] U : ἔκφυσιν Gaza (*exortum*), Heinsius.

THEOPHRASTUS

12.10 οὐκ ἔβλαπτε, πρὸς δὲ τὴν ἕψησιν. ἐν ἑτέροις δὲ καὶ δι' ἄλλα τῆς δυνάμεως οὔσης, οὐκ ἄλογον τὸ μὲν ὑπάρχειν, τὸ δὲ μή · τότε γὰρ ἀβλαστὲς καὶ ἄγονον, ἢ καὶ δυσαυξές, ὅταν κακωθῇ[1] πρὸς τὴν γέννησιν.

οὐκ ἄλογον οὐδὲ ἐξ ἀτεραμόνων τεράμονα[2] γίνεσθαι σπαρέντα πάλιν · ὑπὸ γὰρ τῶν αὐτῶν ἐναντίως διατιθεμένων[3] (ὥσπερ εἴπομεν) οὐκ ἄλογος ἡ τεραμότης. ὥστ' εἰ ὁ ἀὴρ καὶ τὰ ὕδατα καὶ τὰ ἐδάφη τοιαῦτα, τί κωλύει καὶ τὰ σπέρματα μεταβάλλειν;

12.11 εἰ δὲ συμβαίνει τὰ μὲν τεράμονα μᾶλλον ἐνδεδωκέναι, τὰ δὲ ἀτεράμονα περιτετάσθαι τῷ[4] κελύφει (καθάπερ τινές φασιν), καὶ καταχασμωμένων τὰ μὲν μεταβάλλειν τὴν χρόαν, τὰ δὲ μή, καὶ τὰ ἐγχυλότερα[5] θεριζόμενα τεραμονέστερα γίνεσθαι (καὶ γὰρ τοῦτο λέγουσιν), οὐθὲν ἄτοπον · ἢ

---

[1] Schneider : κακανθῇ U.
[2] Heinsius : ἀτεράμονα U.
[3] Gaza (*dispositis*), Schneider : δι (ι from ε [?] U^cc) τιθεμένων U.
[4] τῷ u : το U.
[5] Schneider : ἐκλυτότερα U.

---

[1] Of the bean-plant; powers cannot be observed directly, and must be seen in their results (*CP* 1 21. 4).

DE CAUSIS PLANTARUM IV

to the productivity of the crop, it spoils the crop for cooking purposes. But the power[1] works on different results[2] and for different ends,[3] and so it is not unreasonable that one character should be present, the other not,[4] since only then does the seed fail to sprout and fail to bear (or else grows poorly) when the damage affects its power of reproduction.

12.10

Nor is it unreasonable that stubborn seeds should at the next sowing produce ready ones, since readiness is no unreasonable result of the same causes when they have taken on (as we said)[5] the opposite character; hence, if air, rain and ground are in the opposite state, what is to prevent the seeds from changing too?

If it so happens that ready seeds are more shrunken in the skin, whereas the stubborn ones are fitted tightly by it (as some assert), and if when the pods open the ready seeds change colour whereas the others do not, and if the seeds that are harvested in a more succulent condition turn out to be readier (for this too is asserted), there

12.11

---

[2] Production in the one case of a fertile, in the other of a cookable, bean.

[3] A fertile seed serves the plant, a ready seed serves man.

[4] That is, that the seed should have the character of being fertile, but not that of being ready.

[5] *CP* 4 12. 5.

327

THEOPHRASTUS

τε γὰρ περίτασις[1] σκληρότητος καὶ πήξεως κα ἁδρύνσεως, οὐθὲν δὲ τούτων ἀλλότριον τοῖς ἀτεράμοσιν· τό τε μὴ ἀλλοιοῦσθαι καταπνεόμενα σημαίνει σκληρότητά τινα καὶ ἀπάθειαν, θάτερον[2] δὲ τοὐναντίον· καὶ τὸ μὴ ἔγχυλα[3] θερίζεσθαι, σκληρότητα πλέω καὶ πῆξιν, τὸ δὲ ἄγαν σκληρὸν οὐκ εὐδιάχυτον.

12.12 ὃ καὶ ἐπὶ τῆς ἅλω φασί τινες ἂν συμβῇ, γίνεσθαι ἀτεράμονα,[4] οὐκ ἄγαν τοῦτό γε λέγοντες πιθανόν· ἡ γὰρ ἀπὸ τοῦ ἡλίου ξηρότης ἀχυλότερα μὲν ποιεῖ καὶ ἧττον ἡδέα,[5] εὐεψητότερα <δὲ>[6] οὐδὲν ἧττον, ἀλλὰ μᾶλλον (τὰ γοῦν ἐρίγματα[7] διηλιωθέντα θᾶττον διαχεῖσθαι[8])· εἰ δ' ἄρα καὶ τοῦτ' ἀληθές, οὐκ ἐναντιοῦται τοῖς πρότερον.

οὐδ' εἰ παλαιούμενα γίνεται μᾶλλον ἀτεράμονα· καὶ[9] οὕτως ξηρότερα συμβαίνει γίνεσθαι διά τε τοῦ περιέχοντος καὶ ὑπὸ τῆς διεκπνοῆς

---

[1] Gaza (*obtentio*), Scaliger : περίστασις U.
[2] aP : θατερον (-έ- u N) U.
[3] Schneider : ἔκλυτα U.
[4] Gaza, Scaliger : τεράμονα U.
[5] u : ηδε U.
[6] u.
[7] U : -ή- u.
[8] U : διαχεῖται Schneider.
[9] καὶ <γὰρ> Itali.

is nothing odd. For the tight fit of the skin is a matter of hardness and stiffening and plumpness, and none of these characters is out of place in stubborn seeds; and undergoing no alteration[1] when exposed to the wind points to a certain hardness and immunity to being affected, whereas to undergo it points to the opposite; and not to be harvested in a succulent condition points to a greater hardness and stiffness, and what is very hard is not easily loosened by boiling.

Some say that if this non-succulence also occurs on the threshing floor the seeds become stubborn, but here at least the assertion is not very convincing, for dryness from exposure to the sun makes the seeds less succulent and palatable to be sure, but does not make them the less easy to boil, but easier: thus bruised seeds after being set out in the sun are said to be more quickly softened by cooking. Still, even if this further assertion is true, it does not conflict with what was said before.[2]

12.12

Nor is there a contradiction if seeds kept long in storage tend more to become stubborn: under these conditions too it happens that they get drier, both through the agency of the surrounding air[3] and by the evaporation of their fluid, which takes their heat

---

[1] That is, alteration of colour.
[2] *CP* 4 12. 11 *ad fin.*
[3] Which chills and thereby dries them.

[τοῦ θερμοῦ],[1] ὃ συνεξάγει καὶ τὸ θερμόν.[2]

εὔλογον δὲ καὶ θᾶττον καὶ μᾶλλον κόπτεσθαι τὰ τεράμονα· καὶ γὰρ γλυκύτερα (ταῦτα <δὲ>[3] μᾶλλον ζωοποιεῖ) καὶ πεπεμμένα (μεταβολὴ δὲ καὶ τούτων θάττων).

καὶ ταῦτα μὲν ἂν ἔχοι τὰς εἰρημένας αἰτίας.

12.13 τὸ[4] δὲ μόνα τῶν ὀσπρίων, ἢ καὶ τοῦ παντὸς σίτου, κύαμον καὶ φακὸν ἀτεράμονα γίνεσθαι, ψεῦδος ὑποληπτέον· ἀλλὰ μάλιστα διαπειρώμενοι τούτων διὰ τὴν χρείαν, ταῦτα μόνα φαμέν, ὅτι τοῖς μὲν ἀλήθοντες χρώμεθα, τοῖς δ' ὅλοις. αὕτη δ' οὐ μικρὰ διαφορά· καὶ γὰρ ἐπ' ἐκείνων, εἰ λεπτύνοντες, οὐκ ἂν ὁμοίως ἔνδηλον ἦν. ἐκφαίνεται

---

[1] ego.
[2] U N P^{ac} (?) : ὑγρόν aP^{c}.
[3] Wimmer : <καὶ> Basle ed. of 1541 after Gaza.
[4] U aP : τὰ u N.

---

[1] Cf. CP 1 1. 3.
[2] Cf. CP 4 14. 5; 5 18. 2.
[3] They must change to something worse because they cannot get better; they have as it were reached the end of the line. Cf. CP 5 18. 2.
[4] Cf. HP 8 8. 6: "The terms 'ready' and 'stubborn' are used only of pulses, but it is not unreasonable that the like

along with it.[1]

It is reasonable that the ready seeds among them should also get worm-eaten sooner and to a greater extent, since they are not only sweeter than the rest (and sweet things breed worms more),[2] but they are also concocted (and concocted things too are quicker to change to something else).[3]

These matters too, then, would have the causes mentioned.

Two Notions about Readiness and Stubbornness Are False, and These Causes Do Not Apply

That bean and lentil are the only pulses (or indeed the only grains) that become stubborn we must take to be false. The truth is that we test bean and lentil most of all by our way of using them for food and so assert that they alone get stubborn, because we prepare the rest by grinding but use these seeds entire.[4] This is no small difference, since if we reduced bean and lentil to meal, like the rest, their stubbornness would not be so evident; 12.13

or the same should also occur in cereals, but is not so evident there because we do not prepare them in the same way. Indeed it is not equally apparent in all pulses, but is mainly spoken of in connexion with beans and lentils, whether these are the seeds most affected or whether the effect is more apparent because of our way of preparing them."

THEOPHRASTUS

γοῦν καὶ ἐπὶ τῶν ἐρεβίνθων καὶ ἐπὶ τῶν ἄλλων ἑψομένων ὅλων[1] τὸ ἀτέραμον. εἰ δ' ἔτι μᾶλλον ἐν τούτοις,[2] οὐδὲ τοῦτο ἄλογον · ἀσθενέστερα γάρ (ὡς εἰπεῖν), τὸ δ' ἀσθενὲς παθητικώτερον.

ὃ δὲ λέγουσιν οἱ πολλοί, διότι[3] τὸ κερασβόλον ἀτέραμον γίνεται, μή ποτ' ἄγαν εὔηθες ᾖ · σκληρότερος γὰρ ὁ λίθος, πρὸς ὃν πολλάκις προσπίπτει τὰ σπέρματα, κἂν μὴ προσκόψῃ, μηδὲ βουσὶν ἀροτριᾷ τις, οὐδὲν ἧττον ἀτέραμον γίνεται.

περὶ μὲν οὖν τούτων ἱκανῶς εἰρήσθω.

13.1 ὅσα δὲ τῶν ὁμογενῶν μὴ ἅμα διαβλαστάνει τοῖς ἄλλοις, ἀλλ' ὕστερον πολλῷ (καθάπερ ἐλέχθη περὶ τοῦ τευτλίου), παραπλησία τις ἡ αἰτία ταύτῃ (καθάπερ εἴπομεν) ἐστίν · ὥσπερ γὰρ πρὸς ἕψησιν καὶ ὅλως πύρωσιν ἀτεράμονα, καὶ πρὸς ἔκφυσιν οὕτω καὶ πρὸς βλάστησιν · ἡ γὰρ τοιαύτη, ἀτέλειά τις τῶν σπερμάτων, ὁμοίως δ' ἐπὶ πάντων ὧν τοῦτο συμβαίνει.[4]

---

[1] Schneider : ὅλως U.
[2] δ' ἔτι μᾶλλον ἐν τούτοις ego (ceterum [si quis] illa tantummodo incoctilia [esse contendat] Gaza : δὲ τὸ ἀτέραμον μόνον ἐν τούτοις Schneider : δ' ἀτεράμονα μόνα ταῦτα Wimmer) : δ' ἀτέραμον ἐν τούτούτοις U.
[3] Schneider : διατι U.   [4] U^r N aP : συμβαίνειν U^{ar}.

stubbornness in any case is apparent in chickpea too and the rest when they are boiled whole. If it is still more apparent in bean and lentil, this too is not unreasonable, since these are weaker (one may say) than the rest,[1] and what is weak is more susceptible.

The popular notion that a seed "that strikes the hoof" becomes stubborn[2] is, one fears, much too naive. Thus a stone, and the seeds often collide with them, is harder than a hoof; and even if there is no collision with a hoof, no oxen being used by the farmer in his ploughing, the seed becomes stubborn none the less.

Let this suffice for the discussion of these points.

Stubbornness For Sprouting

In plants of the same kind where some do not come up with the rest, but much later (as we said of beet),[3] the cause is similar to stubbornness: just as some seeds are stubborn about boiling (and cooking in general), so some are stubborn about germination and sprouting, the stubbornness here being a certain immaturity of the seeds (and similarly with all plants in which the delay occurs).

13.1

---

[1] For the weakness of bean *cf. CP* 2 12. 5.
[2] Cf. Plato, *Laws*, ix, 853 D; Plutarch, *Table-Talk*, vii. 2. 1–3 (700 C–701 D); *Geoponica*, ii. 19. 4; xv. 1. 27.
[3] *CP* 4 3. 2; *cf. CP* 2 17. 7; 4 6. 7.

THEOPHRASTUS

περὶ μὲν οὖν τούτων ἐπισκεπτέον.

13.2 περὶ δὲ τοῦ ἰσχυρότερα καὶ εὐχυλότερα γίνεσθαι, καὶ νοστιμώτερα ἢ ἀνοστότερα, καὶ πρὸς τὴν σίτησιν βελτίω ἢ χείρω,

τὰ μὲν τοῖς τόποις διαφέρει (καθάπερ εἴρηται), τῷ ξηρότερα ποιεῖν καὶ πυκνότερα καὶ ξυνεστηκότα μᾶλλον, οἷον τὰ ὀρεινὰ τῶν πεδιεινῶν, καὶ ἔτι μᾶλλον τῶν ἐπόμβρων, εὐπνούστερά τε γάρ, καὶ ἐν τροφῇ συμμετροτέρᾳ.[1]

νοστιμώτερα δ' ἐκ τῆς αὐτῆς παρὰ τὴν τοῦ ἀέρος κρᾶσιν· οὐ γὰρ πάντως ὅταν πλεῖστος καὶ ἁδρότατος γένηται, καὶ νοστιμώτατος, ἀλλ'

---

[1] ego : -οτερα U : -ότερα u.

---

[1] The yield in meal is calculated by comparing the volume of meal or flour with the volume of grain from which it was ground. This is distinct from what we render as "a big yield" or a "small yield," since these are

DE CAUSIS PLANTARUM IV

These matters need investigation.

*Seed Crops: Differences in Consistency,
Yield of Meal and Excellence as Food*

We pass to greater strength and succulence, 13.2
greater or smaller yield in meal[1] and superiority or
inferiority as food.

(1) *Due to the Country*

In some plants the difference is due to the country
(as we said),[2] because the country makes them
drier, of closer texture and more compact, as mountain districts do this more than the plains and still
more than rainy districts, since they are better ventilated and more nearly provide the right amount of
food.

(2) *Due to the Air*

In grain from the same country the greater yield
of meal depends on the tempering of the air, since
it does not necessarily follow, when the crop has
kernels of the greatest number and size, that the
yield in meal is greatest too, but for this a different

---

computed by comparing the amount of grain sown with
the amount harvested and winnowed.

[2] *CP* 3 21. 1.

ἑτέραν δεῖ διάθεσιν ἔχειν.

13.3 δοκεῖ δὲ μεγάλα συμβάλλεσθαι[1] καὶ ἡ σκάλισις πρὸς τὸ νοστιμώτερον ποιεῖν, καὶ τὸ ἐγχυλότερα θερίζειν (ἐξαναλωθέντος γὰρ παντὸς τοῦ ὑγροῦ καὶ χείρω πρὸς τὴν σίτησιν καὶ ἐλάττω). τὰ δ' ὄσπρια θερίζουσιν ἐγχυλότερα καὶ πρὸς τὸ δύνασθαι συλλέγειν, ξηρανθέντα[3] γὰρ καταρρεῖν.[4] τὸν δὲ θέρμον[5] ἤδη λήγοντος τοῦ θέρους[6] [δὲ],[7] ἀνυγραινομένου[8] τοῦ ἀέρος[9] κατὰ τὸ ἑωθινόν· ἂν γὰρ πρότερον θίγῃ τις, ἐκπηδᾷ[10] καὶ οὐκ ἔστι[11] λαβεῖν.

καὶ ταῦτα μὲν ὡς γεωργικά.

13.4 τὰ δὲ τῆς φύσεως, οἷον τὸ ἁδρύνειν τὰ πνεύ-

---

[1] N aP : συμβάλεσθαι U.
[2] Itali (vegetiora Gaza) : ἐγγύτερα U.   [3] B : ξηραθέντα U.
[4] U^c : καταρεῖν U^{ac} : καταρρεῖ Heinsius (defluunt Gaza).
[5] δε θέρμον U^c : δέρμον U^{ac}.   [6] Itali : ἀέρος U.
[7] N aP.   [8] N aP : -μένους (-μενος U^{ac}?) U.
[9] τοῦ ἀέρος N aP omit.
[10] ego (ἐκπίπτει u) : ἐκπῆν (-ν uncertain) U.
[11] u aP : οὔκετη U : οὐκέτι N.

---

[1] *Cf. HP* 8 8. 2: "When there has been a year with good weather the grain also yields more meal. At all events the barley at Athens yields the most meal, Athens being the best barley producer. This happens not when the barley is

DE CAUSIS PLANTARUM IV

condition is required.[1]

(3) *Due to the Farmer*

Hoeing too is held to contribute much to a greater 13.3
yield in meal, and so too the harvesting of the seeds
when they are still juicy, since after all the fluid has
been exhausted they are not only inferior as food but
smaller. Farmers harvest pulse when it is still juicy
also to be able to gather it, since they say that the
seed falls out when dry.[2] Lupine on the other hand
is harvested in the early morning, when summer is
already at an end and the air is getting rainy, for if
you handle it before this time the seed jumps out
and cannot be found.[3]

These are matters for the farmer.

Seed-Crops: Special Effects of Wind and Rain

To the realm of nature on the other hand pertain 13.4

---

most plentiful but when it gets weather with a certain blend of qualities."

[2] *Cf. HP* 8 11. 3: "They harvest legumes when still juicy to collect them more efficiently and easily, for the seeds quickly drop out otherwise and get dry and broken; and they do the same with wheat and a certain kind of barley because they make better meal when not dried out."

[3] *Cf. HP* 8 11. 4: "When left unharvested wheat keeps best, and lupine still more; for they do not even harvest it before there has been rain, because otherwise the seed jumps out when you harvest it and is lost."

THEOPHRASTUS

ματα, καὶ τὸ[1] βόρεια μᾶλλον, καὶ ὅλως οἷς ἕκαστα[2] ψυχρά, καὶ τἀναντία δὴ φθείρειν, καὶ ὅσα δὴ τοῖς ὕδασιν ἢ τοῖς πνεύμασιν ἀπόλλυνται τοῖς παρώροις (ὥσπερ ἐλέχθη περὶ ἐρεβίνθων τῶν ἀνθούντων[3])· ὁ δὲ κύαμος, ἐὰν πνεῦμα ἐπιγένηται λαμπρόν· καὶ ὁ πυρὸς καὶ ἡ κριθή, κἂν[4] ἀπηνθηκότα, ὑγρὰ δ' ἔτι[5] ληφθῇ, διαπνεῖται γάρ, καὶ κοπτόμενα πρὸς ἄλληλα κενοῦται.

τελεουμένων[6] δὲ ἐπιγινόμενον ὕδωρ, ἐναντίως τὴν μὲν κριθὴν βλάπτει, τὸν δὲ πυρὸν ὠφελεῖ μᾶλλον· καὶ <γὰρ>[7] γυμνή, καὶ τὸ ὅλον ἀσθενής, ὁ δὲ καὶ ἐν χιτῶσιν, καὶ πυκνότερον καὶ ἰσχυρότερον, ὥστε τὴν μὲν ὀλίγης δεῖσθαι τροφῆς (καὶ σχεδὸν ἀπὸ[8] τοῦ ἀέρος μόνης), τὸν δὲ πλείονος· ἔτι δέ,[9] ἰσχυρότερος ὤν, καὶ κατακρατεῖ καὶ συμπέτ-

13.5

---

[1] U : τὰ Schneider.
[2] οἷς ἕκαστα U : ἃ ἑκάστοις Wimmer.
[3] N aP : ἀθοῦντων (-οὐ- u) U.
[4] ego : καὶ U.
[5] ego (δ' ἔτι ἂν Wimmer) : δέ τινα U.
[6] Wimmer : κενουμένων U^c (-οῦ- U^{ac}).
[7] καὶ <γὰρ> Wimmer (ἡ <μὲν γὰρ> Itali) : καὶ U.
[8] U : τῆς ἀπὸ Schneider.
[9] ego (ἐκπίπτει U : ἐκπῆν (-ν uncertain) U.

---

[1] CP 4 8. 4; cf. HP 8 6. 5.

such matters as these: that winds make the seeds plump, and northerly winds do so more than the rest, and so in general whatever regional winds are cold, the winds of the opposite kind destroying them; and the cases where the crop is lost from unseasonable rain or wind (as was said[1] about the chickpea in flower). The bean crop is lost if a strong and steady wind comes up while the bean is in flower, wheat and barley even after shedding their flower if the wind catches them while the seeds are still fluid,[2] for the fluid evaporates and the plants are shaken empty of their grain by being buffeted against one another.

When rain comes as the seeds are reaching maturity it has opposite effects on wheat and barley: it injures barley but helps wheat instead. For barley is naked and in general weak, but wheat has several coats and is closer in texture and stronger; so that barley needs little food, indeed (one might say) only what comes from the air, whereas wheat needs more. Furthermore, since wheat is stronger, it succeeds better not only in mastering the food[3] but

13.5

---

[2] *Cf. HP* 8 10. 3: "Both wheat and barley are also killed by winds, when caught either in flower or just after they have shed the flower and are weak, and barley more especially, and often when it is already getting plump, if great winds arise and last for some time, for they dry and wither it, and some call this getting 'wind-blown.'"

[3] That is, the rain water.

THEOPHRASTUS

τει μᾶλλον, καὶ τὸ ὅλον οὐδὲ πολλὴν δέχεται διὰ τὴν πυκνότητα καὶ τοὺς χιτῶνας·[1] ἡ δέ, πλείω τε ἕλκει, μανὴ τὴν φύσιν οὖσα,[2] καὶ ταύτην οὐ καταπέττει δι' ἀσθένειαν, καὶ ὑγρᾶς γενομένης, ὁ ἥλιος ἅμα τῇ ἐπιγενομένῃ συνεξάγει τὴν οἰκείαν ὑγρότητα[3] συνεφελκομένην,[4] καὶ τὰ πνεύματα δὲ κόπτοντα φθείρει μᾶλλον διὰ τὴν ἀσθένειαν.

13.6 ἀπορεῖται δὲ καὶ διὰ τί, ἁδροῦ ὄντος τοῦ σίτου (καὶ σχεδὸν ὥσπερ ξηροῦ), ἐφύσαντος,[5] οὐχ ὅτι βελτίων, ἀλλὰ καὶ χείρων[6] γίνεται· ἐὰν δὲ θερισθεὶς εἰς θωμοὺς[7] συντεθῇ, ἁδρότερος καὶ βελτίων, ἔνιοι δὲ καὶ ῥαίνουσιν.

αἴτιον δ' ὅτι τότε[8] μὲν ἀνυγραίνεται καὶ ὁ ἥλιος[9] ὅταν ἀναλάμψῃ συνεξάγει τὴν οἰκείαν

---

[1] Scaliger : χειμῶνας U.
[2] U^r N aP : οὖσαν U^{ar}.
[3] U^{cc} (-τη U^{ac}).
[4] Heinsius : -η U.
[5] ego (ἐφυσθεὶς Wimmer : ὅταν ἐφύσῃ Schneider) : ἔκφυσις U.
[6] u aP : χεῖρω U (-εί N).
[7] Heinsius : θωσμ- U.
[8] Gaza, Schneider : τὸ U.
[9] ὁ ἥλιος u : ὅλιος U.

in concocting it, and indeed does not even absorb much of it because of its close texture and coats; whereas barley absorbs it in a greater amount, being in its nature of open texture, and is too weak to concoct properly what it absorbs; and when it has thus become fluid, the sun extracts together with the added fluid the native fluid, which is drawn away with it; and again, the winds by their buffeting are more destructive to it because of its weakness.

A Problem about the Mature Wheat and Barley

Another problem[1] is this: why, when the cereal is plump and one might say dry (as it were),[2] rain, far from improving the crop, makes it worse; but cereal that has been reaped and heaped into piles gets plumper and so improves, and some farmers even sprinkle the pile?

Solution

The reason is this: in the first case the cereal gets drenched and the sun comes out again and removes

---

[1] The first problem (though not called one) was dealt with in the second paragraph of *CP* 4 13. 4 and the chapter following.

[2] After shedding the flower wheat and barley were "fluid" (not firm or rigid): *CP* 4 13. 4, first paragraph *ad fin.*

THEOPHRASTUS

ὑγρότητα καὶ ἰσχναίνει·[1] ὅταν δὲ εἰς θωμοὺς[2] συντεθῇ, συνικμάζεταί τε, καὶ ἡ ἀναγομένη ἀτμίς, λεπτὴ καὶ πνευματώδης οὖσα, παρεισδύεται καὶ ἁδρύνει τοὺς ὄγκους.

13.7 ταὐτὸ δὲ τοῦτο συμβαίνει καὶ ὅταν εἰς τὰ οἰκία[3] τεθῇ χύδην, διὸ καὶ τὸ ἐπίμετρον ποιεῖ· τὸν γὰρ ἀτμὸν τὸν ἀνιόντα, λεπτὸν ὄντα, δέχεται, καὶ διὰ τοῦτο εἰς βάθος καταβάλλουσιν, ὅπως πλέον ἀνίῃ.[4] πάντα δὲ ταῦτα διασημαίνει, καὶ ἔτι πρὸς τούτοις ἡ ἐμβαλλομένη γῆ καὶ ἄκοπον παρέχουσα καὶ ἀνοιδίσκουσα, καὶ ὅτι δύναταί τινα δι᾽ αὐτῶν ἕλκειν[5] τροφὴν ἄνευ τῶν ῥιζῶν, ὅθεν καὶ ἡ ὑπὸ τῶν πνευμάτων καὶ τοῦ ἀέρος οὐκ ἄλογος.

---

§ 7 lines 2–7 *Cf.* Plutarch, *Quaest. Conv.* v. 5. 1 (676 B) (after a reference to Theophrastus): ἔτι δὲ καὶ καταμιγνυμένη (*sc.* ἄργιλος) πρὸς σῖτον ἐπίμετρον ποιεῖ δαψιλές, ἁδρύνουσα καὶ διογκοῦσα τῇ θερμότητι τὸν πυρόν.

---

[1] Scaliger : ἰσχάνει U.
[2] v : θωσμοὺς U N aP.
[3] u P : οἰκεία U : οἰκεῖα N a.
[4] ἀνίῃ u : ἂν εἴη U N aP.
[5] u : ἕλκει U.

the native fluid with the rain and so shrinks the kernels; in the second case dampness is produced in the pile and the vapor that arises, which is thin and like *pneuma*,[1] penetrates the kernels and makes them plumper.[2]

The same thing also happens when the grain is 13.7 dumped in store rooms, and this is why it gives "good measure": it absorbs the vapour that arises, which is thin, and for this reason the grain is piled deep, to let the vapour rise further. All this and the further circumstance that earth[3] thrown on the grain keeps it from getting worm-eaten[4] and makes it swell indicates that the kernels can also attract a certain amount of food by themselves, without their roots, and hence their also being fed by the winds and the air is not unreasonable.

---

[1] Warm and expansive gas.

[2] *Cf. HP* 8 11. 4: "This is why they heap both wheat and barley in piles, and they are considered to get plumper in a pile and left unwinnowed."

[3] In this connexion Theophrastus (according to Plutarch, *Quaest. Conv.* v. 5. 1 [676 B]) speaks of clay: "furthermore when it (*sc.* clay) is mixed in with grain it produces abundant 'good measure,' making the wheat plump and bulky by its heat."

[4] *Cf. HP* 8 11. 7: "There is also held to be a kind of earth in certain countries that when sprinkled on the wheat preserves it, as the earth at Olynthus and at Cerinthus in Euboea . . ."

THEOPHRASTUS

14.1 τὸ¹ δὲ τῆς ἐρυσίβης κοινὸν οὐχ ἧττον, ἀλλὰ μᾶλλον ἅπτεται τῶν σιτωδῶν, κριθῆς δὲ καὶ μᾶλλον καὶ² πυροῦ, διά τε τὸ γυμνοτέραν³ εἶναι (τὸν δ᾽ ἐν⁴ χιτῶσι πλείοσιν), καὶ διὰ τὸ ἐγγυτέρω τοῦ στάχυος ἔχειν τὸ φύλλον ὅθεν ἡ ἀπόχυσις, ἐν ᾧ μένει καὶ ἡ ὑγρότης, ὥστε σαπεῖσα μᾶλλον ἅπτεται· καὶ διὰ τὸ τὸν⁵ στάχυν ὀρθὸν εἶναι καὶ πυκνότερον, ἀπορρεῖ γὰρ ἧττον (διὸ καὶ ἐπικύπτειν ξυμφέρει), καὶ ἀπολλύει τὸ συνεχὲς θᾶττον (ἐν δὲ τῷ μανῷ,⁶ πρὸς τοῖς ἄλλοις, καὶ διεκπίπτει ἡ ὑγρότης). ἡ αὐτὴ δ᾽ αἰτία καὶ τῇ λευκῇ⁷ πρὸς τὰς ἄλλας· ἅπαντα γὰρ μάλιστα ἔχει τῶν ἄλλων.

14.2 τῶν δ᾽ ὀσπρίων μάλιστα ἐρυσιβᾷ κύαμος, καὶ διὰ τὸ πολύφυλλος εἶναι πολλαχόθεν, καὶ διὰ τὸ πυκνοσπορεῖσθαι, καὶ διὰ τὸ τὴν ὑγρότητα μά-

---

¹ Schneider : τα U.
² U : ἢ Schneider.
³ Wimmer (τὴν μὲν γυμνοτέραν Schneider) : γυμνοτερα U.
⁴ δ᾽ ἐν Wimmer : δε U.
⁵ διὰ τὸ τὸν Schneider (*quod* Gaza : ἔτι τὸν Wimmer) : ἄτοπον U.
⁶ Wimmer : ἄνω U.
⁷ λευκη U : Ἀχιλληΐδι (?) ego.

---

[1] *Cf. HP* 8 10. 1–2: "Of diseases of seed-crops some are common to all, as rust ... Broadly speaking cereals are

DE CAUSIS PLANTARUM IV

Seed-Crops. Diseases: (1) Rust

Rust is no less common to the rest but attacks cereals more. Further it attacks barley more than wheat[1] for several reasons: barley is more naked (whereas wheat has several coats); barley leaf is closer to the place in the ear from which it heads, and the leaf is where the rain water remains, so that when the fluid decomposes it is better able to infect this part; the barley ear is (1) more erect and (2) of closer texture, so (1) the water does not run off so easily (which is why it is good for the ear to bend),[2] and (2) the disease destroys more quickly what is of continuous texture, whereas open texture, besides its other advantages,[3] lets the water through. The same cause that makes barley more susceptible than wheat makes white barley more susceptible than the other kinds, since it has all these characters to a greater extent than they do. 14.1

Of pulses bean gets rust most, both because it has many leaves coming from many parts, and because it is sown thick, and because it most of all 14.2

---

more liable to rust than pulses, and among cereals barley more than wheat; and among barleys some kinds more than others, and most of all (one might say) Achilles barley."

[2] *Cf. CP* 3 22. 1.

[3] It is not so quickly destroyed and is (*cf. CP* 1 8. 2) in general conducive to growth and feeding.

THEOPHRASTUS

λισθ' ἕλκειν εἰς ἑαυτὸν διὰ τὴν μάνωσιν, καὶ ἔτι διὰ τὸ πρὸς τῇ γῇ μάλιστα τὸν καρπὸν ἔχειν (σήπεται γὰρ μάλιστα τὰ κάτω διὰ τὴν ἄπνοιαν). καὶ ὅλως δὲ τῶν χεδροπῶν[1] τὰ τοιαῦτα.

14.3 ἐρυσίβη <δὲ>[2] σηψίς τίς ἐστι τοῦ ἐφισταμένου ὑγροῦ, διὸ πολλῷ μὲν ὕσαντος οὐ γίνεται (καταπλύνεται γάρ), ἐὰν δὲ ψεκάδες ᾖ καὶ δρόσοι πλείους γένωνται, καὶ ὁ ἥλιος ἐπιλάβῃ καὶ ἄπνοια, τότε σήπεται (διὸ καὶ ἐν τοῖς εὔπνοις καὶ μετεώροις ἧττον, ἐν δὲ τοῖς κοίλοις καὶ δροσοβόλοις μᾶλλον). καὶ πανσελήνοις δὲ μᾶλλον, ὅτι συνεργεῖ καὶ ἡ τῆς σελήνης θερμότης, καὶ ὅλως ὁ ἀὴρ ὑγρότερος.

14.4 πάντων δ' ἐπικηρότατον ὁ πίσος, πρὸς μὲν τὰς ἐρυσίβας, ὅτι πολύφυλλον καὶ χαμαισχιδὲς καὶ εὐαξές (συμπληροῖ γὰρ τὸν τόπον, κἂν ἀραιὸς ᾖ)· πρὸς δὲ τὰ ψύχη καὶ τοὺς πάγους, ὅτι ἀσθενόρριζον.

σκωληκοῦται δὲ μάλιστα <ἢ>[3] μόνα πυρὸς καὶ

---

[1] u : γεδροπῶν U. [2] aP.
[3] Schneider.

---

[1] *Cf. CP* 3 22. 2.
[2] *Cf. HP* 8 10. 2, cited in note 2 on *CP* 3 22. 1.

draws the water into itself because of its open texture, and furthermore because it has its fruit closest to the ground (for the lower parts of a plant get rust the most because of the lack of ventilation). And this holds of all legumes in general that have these characters.

Rust is a kind of decomposition[1] of the fluid that collects on the surface. For this reason it does not occur after heavy rain, since the fluid washes off; but if there have been drizzles and heavy dews followed by sunshine and no wind, decomposition takes place (which is why there is less rust in well ventilated[1] and high ground, more in hollows and where there is dew). Again there is more at the full moon,[2] because the heat of the moon also contributes[1] and the air is in general moister.

14.3

Of all seed-crops the pea is the most delicate. It is susceptible to rust because it is many-leaved, branches out at the ground, and grows well (filling out the space round it even when sown thin); and it is susceptible to cold spells and frost because it has weak roots.

14.4

Seed-Crops. Diseases: (2) Grubs

It is wheat[3] and chickpea that only or mainly

---

[3] Cf. HP 8 10. 4: "Wheat is also destroyed by the grubs, one set as soon as produced devouring the roots ..."; cf. also CP 3 22. 4.

ἐρέβινθος,[1] οὐκ ἐν τοῖς αὐτοῖς δὲ ἑκάτερος, ἀλλ' ὁ μὲν ἐν τῷ καρπῷ, ὁ δὲ ἐν ταῖς ῥίζαις, ἄμφω δὲ διὰ γλυκύτητα · ὁ μὲν ἐρέβινθος, ὅταν[2] ἡ ἅλμη περιπλυθῇ[3] (καθάπερ εἴρηται), ὁ δὲ πυρός, ὅταν ἡ ῥίζα ὑγρανθῇ. ζῳοποιεῖ δὲ σηπόμενα τὰ γλυκέα, γλυκύτερον γὰρ ὁ πυρὸς κριθῆς (διὸ καὶ τὸ ἄχυρον ἥδιον) · ὁ δὲ ἀπόλλυται[4] γινόμενος[5] <ἢ>[6] ὅταν ἐξαναλώσῃ τὴν ἐν τῷ καλάμῳ τροφὴν ἅτερος[7] [ἢ][7] ὥστε ὅλον ἐξαπολλύναι τὸν στάχυν ἢ κατὰ θάτερον μέρος.

ταῦτα μὲν οὖν καὶ τὰ τοιαῦτα καθάπερ νοσήματος ἔχει χώραν, ὑπὲρ ὧν οὐ χαλεπὸν τὰς αἰτίας ἰδεῖν.

15.1 περὶ δὲ τῶν θερινῶν σπερμάτων (οἷον σησάμου,

---

[1] U r N aP : ἐρέμινθος U ar.
[2] Scaliger after Gaza : ἔστ' ἂν U.
[3] Gaza, Schneider : περιπαυθῇ U.
[4] ἀπόλλυται ego : σκώληξ U c (from -ιξ from -ηξ).
[5] U : γενόμενος Scaliger.
[6] ego.
[7] ego : αὐτός U.

---

[1] *Cf. HP* 8 1. 5 and *CP* 3 22. 3; in both passages Theophrastus speaks of "caterpillars." *Cf.* also *HP* 8 10. 1: "The diseases of seed-crops are some common to all, as rust,

get grubs.[1] But the two do not get them in the same parts, chickpea getting them in its fruit, wheat getting them in its roots, and both getting them because of their sweetness; chickpea when the brine is washed off (as we said),[2] wheat when the root gets wet.[3] Sweet things as they decompose breed 14.5 animals, for wheat is sweeter than barley (which is why its bran is more palatable). The wheat perishes in the process of birth or when the second grub has consumed the food in the haulm so as to destroy the whole ear or one side of it.[4]

These occurrences and the like count (as it were) as diseases. It is not hard to see their causes.

Summer Seeds: Their Weakness and Rapid Growth

As for summer seeds,[5] such as sesame, hedge- 15.1

---

some peculiar to certain plants, as gangrene in chickpea and being devoured by caterpillars..."

[2] *CP* 3 22. 3; 4 8. 4.   [3] *Cf. CP* 3 22. 4.

[4] *Cf. CP* 3 22. 4 and *HP* 8 10. 4: "Wheat is also destroyed by the grubs, one devouring its roots as soon as the wheat starts to grow, and the other occurs when the wheat, owing to drought, is unable to head, for the grub is produced in it then and eats the haulm as it is played out. It eats it as far as the ear, and after consuming it perishes. And if it devours the whole haulm the wheat itself perishes, but if it eats one side of the haulm and the wheat manages to head, this part of the ear is withered and the other sound."

[5] *Cf. HP* 8 1. 1, 4 for summer seeds.

THEOPHRASTUS

ἐρυσίμου, κέγχρου, ἐλύμων) ἐπὶ τοσοῦτον μὲν εἰπεῖν ἐστι[1] παντὶ πρόχειρον, ὅτι δι᾽ ἀσθένειαν ταύτην τὴν ὥραν σπείρεται, καὶ διὰ τὸ ταχὺ τελεοῦν· ἡ δὲ φύσις αὐτῶν, ποία τις ἂν εἴη, σκεπτέον. οὐδὲ γὰρ γεώδης ὥσπερ τῶν ὀσπρίων, οὔτε πάντῃ κούφη, τὸ γὰρ σήσαμον λιπαρὸν καὶ δεόμενον πέψεως. καὶ τοῦτο μὲν μονόρριζον· ὁ δὲ κέγχρος καὶ πολύρριζον[2] καὶ βαθύρριζον, καὶ πολυκάλαμον, ὥστε, εἰς ἀμφότερα μεριζομένης τῆς τροφῆς καὶ τῆς δυνάμεως, χρονιώτερον ἐχρῆν εἶναι· τὸ γὰρ σήσαμον, ἐπεὶ μονόρριζον καὶ βαθύρριζον, ἄνω πᾶσαν ἀφίησιν τὴν δύναμιν. ἀλλ᾽ ἀντίκειται τούτῳ τὸ λιπαρόν· τὰ γὰρ αὖ γλίσχρα πλείονος πέψεως δεῖται, ταῦτα δ᾽ (ὥσπερ εἴρηται) πάντα πολύκαρπα.

περὶ μὲν οὖν τούτων σκεπτέον.

15.2 ἐπεὶ δ᾽ ὅλως εἰπεῖν τὰ μικρόκαρπα πολυκαρπό-

---

[1] Gaza, Itali : ὅτι U.
[2] Wimmer : πολύριζος Uᶜ from -υρίζος.

mustard, millet and Italian millet, anyone can readily say this much: they are sown at that season because of their weakness and rapid maturing. But of what description their nature is requires investigation. For it is not earthy either, as in pulse,[1] nor yet entirely light, for sesame is oily, and to be oily it requires concoction. Again, whereas sesame is single-rooted, millet is not only many-rooted and deep-rooted, but also many-haulmed,[2] and in consequence, since its food and power are apportioned both upward and downward, the plant should take longer to mature than it does, since sesame, being single-rooted and deep-rooted, directs all its power upwards. But to this is opposed its oiliness, for viscous things in turn require more concoction, and all these plants (as we said)[3] have an abundant crop.

These matters, then, require investigation.

Summer Seeds: Their Large Yield

Since on the whole plants with small fruit pro-   15.2

---

[1] So legumes make heavier food than cereals (since they are earthy) and yet spend less time in the ground (*CP* 4 9. 1).

[2] *Cf. HP* 8 9. 3: "Of summer seeds sesame is held to be hardest on the soil and to exhaust it most; yet millet has more and thicker haulms and more roots."

[3] *CP* 2 12. 1.

THEOPHRASTUS

τερα, καθάπερ τά τε νῦν εἰρημένα, καὶ οἱ φακοὶ δὴ καὶ ὅσα τοιαῦτα, καὶ ἐπὶ πᾶσι τὰ πολυχούστατα λεγόμενα, κύμινόν τε καὶ μήκων, καὶ ὅλως δὴ τὰ λαχανώδη πάντα καὶ τὰ ἐναγγειόσπερμα πολύχοα, τίς ἂν οὖν αἰτία τούτων εἴη; πότερον ὅτι τὰ μικρὰ ῥᾷον ποιῆσαι; καὶ γὰρ ζῷά φαμεν ἔνια τοιαῦτα, καὶ μάλιστα δὴ τὰ ᾠοτόκα καὶ σκωληκοτόκα.

ἢ αὕτη μὲν ἔξωθεν, ἄλλην δέ τιν'[1] οἰκειοτέραν ζητητέον; ἐπεὶ καὶ τῇδε διαφέρει τῶν περὶ τὰ ζῷα· τῶν μὲν γὰρ ἐκτρέφει τὰ ᾠὰ καὶ τελειοῖ, καὶ ὅλως ζῳογονεῖ, τὸ περιέχον (ὥσπερ καὶ ἐν τοῖς πρότερον εἴρηται)· τῶν δὲ καρπῶν ἐν αὐτοῖς τοῖς φυτοῖς ἡ ἐκτροφὴ καὶ τελείωσις, ὥστε πλείονος ἀεὶ[2] δυνάμεως.

15.3 πρὸς δὲ θησαυρισμὸν ἄριστα μὲν κέγχρος καὶ

---

[1] τιν Uac : τινα Uc.
[2] U : δεῖ Scaliger.

---

[1] *Cf. CP* 2 12. 1–6 and *HP* 8 3. 4–5: "In general legumes have more fruit and the greater yield, and still more than these the summer seeds millet and sesame, and among legumes lentil most of all. Put broadly, one might say that the plants with smaller seeds produce

duce it in abundance,[1] as the summer seeds just now mentioned,[2] lentils and the like, and last but not least the plants that are said to have the biggest yield, cummin and poppy; and since in general all vegetables and all plants with seeds in capsules have a large yield—what would be the reason for this? Is the reason that it is easier to produce what is small? Certain animals too, we say, bring forth a small and numerous progeny, especially the oviparous and larviparous.[3]

Or is this cause one that is found outside plants, and we should look for another belonging more especially to them? Since what happens with the plants differs also as follows from what happens with the animals: it is the surroundings[4] that rear the eggs of the animals and mature them and in a word make animals of them (as we also said earlier),[5] whereas the rearing and maturing of the fruit takes place in the plants themselves, and so always involves more power.

Seed-Crops that Keep in Storage

Millet, Italian millet, sesame, lupine and chick- 15.3

---

more of them, as cummin among vegetables, though all produce many."   [2] *CP* 4 15. 1.

[3] *Cf.* Aristotle, *On the Generation of Animals*, iv. 4 (771 a 17–b 14).   [4] That is, the air or water.

[5] *CP* 4 6. 4.

THEOPHRASTUS

ἔλυμος καὶ σήσαμον καὶ θέρμος καὶ ἐρέβινθος, τὰ μὲν ὅτι πολυχίτωνα[1] καὶ ξηρά, τὸ δ' ὅτι λιπαρόν, ὁ δὲ θέρμος καὶ ὁ ἐρέβινθος[2] ὅτι πικρότητά τινα ἔχουσι καὶ δριμύτητα, διατηροῦσι δ' αὗται[3] τὸ ἀπαθές (ὥσπερ εἴρηται), ὅθεν καὶ διαφθειρόμενα <μόνα>[4] τῶν σιτωδῶν οὐ ζῳοῦται (καθάπερ οὐδὲ τὰ τῶν λαχάνων· καὶ γὰρ ταῦτα ἄζῳα ἢ ἀπαθῆ[5] τῇ ξηρότητι ἢ τῇ δριμύτητι), συμβαίνει γὰρ ὥσπερ ἀνάμιξίν τινα γίνεσθαι τοῦ ἔξωθεν.

σκεπόμενον[6] δὲ ἕκαστον ἰδιά τινα ποιεῖ τὴν[7] μορφὴν[8] ἐκ τῆς οἰκείας ὑγρότητος, οἷον οἱ μὲν πυροὶ καὶ αἱ κριθαὶ τοὺς κίας, οἱ δὲ κύαμοι τὸν ὑπό τινων καλούμενον μίδαν· ὡσαύτως δὲ καὶ οἱ φακοὶ καὶ αἱ ἀφάκαι καὶ πίσοι καὶ ἄλλα. τὸ δ' αἴτιον κοινὸν πλειόνων, οὐ γὰρ μόνον ἐν τοῖς καρποῖς,

---

[1] U^c from πολυτωνα.
[2] ὁ ἐ. u : ὁρέβινθος U^{cc} from -ν.
[3] Gaza, Schneider : αὗται U.
[4] Schneider.
[5] ego : ὅσα δὴ U.
[6] U : σηπόμενον Itali.
[7] ego : ἰδίαν τινὰ ποιεῖται U.
[8] Dalecampius (*animal* Gaza) : τροφὴν U.

---

[1] *Cf. CP* 4 2. 2.
[2] *CP* 4 2. 2.

pea keep best in storage,[1] the first two because they are many-coated and dry, sesame because it is oily, lupine and chickpea because they have a certain bitterness and pungency. Bitterness and pungency, as we have said,[2] preserve things from being affected, and for this reason these alone of seed-crops do not on decomposing breed animals. No more do the seeds of vegetables, for these too by reason of their dryness or pungency breed no animals (or are unaffected by the decomposition that breeds them, since what happens instead is a kind of intermingling with them of what is outside them).

15.4

Seed-Crops: On Getting Worm-Eaten in Storage

Each kind of seed-crop when put under a roof produces from its proper fluidity certain animals of a form peculiar to itself: so wheat and barley produce their weevils, beans the creature that some call "midas,"[3] and so with lentils, tares, peas and others.[4] The cause is common to a number of things, for it applies not only to seed-crops, but also to

---

[3] Perhaps the movement of its antennae resembled that of an ass's ears.

[4] *Cf. HP* 8 11. 2: "There arise in seed-crops as they decompose creatures peculiar to each, as we said (*cf. ibid.* 8 10. 5), except in chickpea, for chickpea is the only seed-crop not to breed animals. And all things that decompose can breed a grub, but when seeds get worm-eaten each breeds a creature of its own."

THEOPHRASTUS

ἀλλὰ καὶ ἐν τοῖς δένδροις καὶ μετὰ ταῦτα τοῖς ξύλοις · καὶ ὅλως ὅσα τῶν ἀψύχων ζῳογονεῖται διαφόρους ποιεῖ πάσας τὰς μορφάς, ὡσὰν ἐξ ἑτέρας ὕλης.

16.1 ὁ δὲ κονιορτώδης σῖτος θᾶττον σήπεται, καὶ ὁ ἐν τοῖς κονιατοῖς ἢ ἀκονιάτοις[1] οἰκήμασιν,[2] διὰ τὸ πλείω θερμότητα ἔχειν (καὶ γὰρ [ᾖ][3] κονιορτὸς θερμός, ἅτε ξηρὸς ὤν, καὶ τὸ κόνισμα, ὅταν ἀναδέξηται, τηρεῖ τὴν θερμότητα) · ἐπεὶ καὶ ἀθρόος καὶ κατὰ βάθους ὤν, πλείων γὰρ ἡ θερμότης. θᾶττον δὲ πολὺ κόπτεται κριθὴ πυροῦ διὰ τὸ μανότερον εἶναι καὶ γυμνότερον · τάχιστα δ' ὁ κύαμος, θερισθεὶς γὰρ εὐθὺς καὶ συντεθεὶς ἐνιαχοῦ, καθάπερ ἐν
16.2 Θετταλίᾳ. φαίνεται δὲ κατὰ τοὺς τόπους · παρ' ἐνίοις γὰρ ἀσηπτότατος (ὥσπερ ἐν ταῖς ἱστορίαις εἴρηται περὶ Ἀπολλωνίαν · περὶ[4] Τάραντα καὶ ἀναπληροῦσθαι δέ φασι κοπέντα). τοῦτο μὲν οὖν σκεπτέον.

---

[1] Gaza (*non levigatis*), Itali : κονιατοῖς U.
[2] u : ἢ κήμασιν U.
[3] ego : ᾖ from ἢ U^c (?) : ἡ N : ὁ aP.
[4] <καὶ> περὶ Schneider.

---

[1] *Cf. CP* 3 22. 3–6; 5 10. 5.
[2] *Cf. HP* 5 4. 4–5.   [3] *Cf. HP* 8 11. 1.
[4] *Cf. HP* 8 11. 3: "It appears that one country and one

DE CAUSIS PLANTARUM IV

trees[1] and to the wood[2] that is later derived from them; and in general all inanimate things in which animals are bred produce different shapes in each case, since each animal comes from a different matter.

Dusty cereals decompose faster, and cereals stored in plastered faster than in unplastered rooms[3] because in these cases the cereals have more heat (for not only is dust hot, since it is dry, but plaster retains heat that it has once absorbed); indeed cereals also decompose faster when piled deep, since their heat is then greater. Barleycorns get worm-eaten much faster than wheat kernels because the barleycorn is more open in texture and more naked. Bean gets worm-eaten fastest, for in some countries (as in Thessaly) this happens directly after harvesting and storage. This appears to depend on the district, for in some places the bean is very free from decomposition, as at Apollonia, as we said in our History[4]; and in Tarentum it is said that the bean after getting worm-eaten can even regenerate the lost part[5]—a matter we must investigate.

16.1

16.2

kind of weather differs from another in the matter of whether seed-crops get worm-eaten or not. Thus they assert that at Apollonia on the Ionian Sea beans never get worm-eaten at all, and are therefore stored away; and beans keep at Cyzicus for a considerable time."

[5] Not mentioned elsewhere.

THEOPHRASTUS

ἀκόπως δὲ ἔχει τάδε · θέρμος κέγχρος αἰγίλωψ αἶρα σήσαμον ἐρέβινθος (καὶ ὅλως τὰ δριμέα). ὁ μὲν οὖν αἰγίλωψ πολυχίτων, ἡ δ' αἶρα γυμνή · διὸ καὶ ἄλλῃ τινὶ δυνάμει. τὸ δ' ὅλον ἄκοπα τὰ μὲν ξηρότητι (διὸ καὶ ἡ κάχρυς χρονιώτερον), τὰ δὲ ἀμιξίᾳ καὶ καθαρότητι[1] (καθάπερ ὁ χόνδρος), τὰ μὲν[2] χυμῶν φύσει καὶ δυνάμει φυσικῇ (καθάπερ ἐρέβινθος θέρμος σήσαμον), τὰ δὲ τῷ πολὺ τὸ ἀποστέγον ἔχειν (ὥσπερ ὁ ὦχρος · ἅμα δὲ τοῦτό γε καὶ ξηρὸν καὶ θερμόν). τῷ δὲ σίτῳ[3] κοπέντι βοήθεια ταχίστη τὸ εἰς τὴν ἅλω φέροντας ἀπικμῆσαι.

καὶ τούτων μὲν σχεδὸν ἐνταῦθά που τὸ αἴτιον.

16.3   ὃ δὲ παρά τινων ζητεῖται, διὰ τί ποτε ὁ κεκομμένος σῖτος καὶ παλαιός, ὕδατος μὲν ἐπιχεομένου θερμοῦ, διαβλαστάνει (ταύτην[4] γὰρ δὴ διάπειραν λαμβάνουσιν εἰ φύσιμος, ὅταν ἐμβάλλοντες[5] εἰς σκάφιον ἐπιχέωσιν), σηπόμενος δ' οὐ διαβλαστάνει — καίτοι καὶ ἡ σῆψις ὑπὸ θερμοῦ, καὶ ἀνυ-

---

[1] aP^c : καθαρόν τι U N P^{ac} (?).
[2] U : δὲ Itali.
[3] δὲ σίτῳ Scaliger (*frumento* Gaza : σίτῳ Itali) : ἀσίτω U.
[4] U^{ac} : -ηι U^c.   [5] U^{cc} (a from λ) : ἐμβαλόντες Schneider.

DE CAUSIS PLANTARUM IV

The following do not get worm-eaten: lupine, millet, haver-grass, darnel, sesame, chickpea (and all pungent seeds in general). Now aegilops has many coasts, darnel none; darnel therefore has some other power of resistance. In general some seeds escape getting worm-eaten by reason of their dryness (which is why parched barley keeps longer), whereas others do so by staying clear of mixture and remaining pure (like groats), some doing this by the nature of their flavour and their natural[1] power, as chickpea, lupine and sesame, others by their thick insulation against the outside, as bird's pease (but this seed is at the same time also dry and hot). When cereal gets worm-eaten the quickest remedy is to take it to the threshing-floor and winnow it.

The causation of all this, then, lies (I may say) somewhere here.

An Easy Question about Worm-Eaten Cereal

The question raised by some, why old and worm-eaten cereal sprouts when warm water is poured on it (this being the test that people use to decide whether it can germinate), whereas decomposing cereal does not, although decomposition too is produced by heat and involves liquefaction, would

---

[1] Internal.

THEOPHRASTUS

γραίνεται — κομιδῇ δόξειεν ἂν παρ' ὁμωνυμίαν[1] εἶναι. διαφέρει γὰρ ἥ τε φθαρτική, καὶ ἡ γόνιμος, θερμότης, καὶ ἡ ὑγρότης δ' ὡσαύτως· ἔστι δ' ἡ μὲν σήπουσα ἀλλοτρία καὶ φθαρτική, ἡ δὲ[2] τοῦ ὕδατος τοῦ θερμοῦ, συμμετρίαν τινὰ ἔχουσα (δεῖ-

16.4 ται γὰρ τοιαύτης πρὸς τὴν βλάστησιν). ἐπεὶ τρόπον τινὰ παρόμοιον καὶ εἴ τις ἀπορoίη διὰ τί ἡ θερμότης καὶ ἡ ὑγρότης καὶ ὑγίειαν ἐν τοῖς σώμασι ποιοῦσι[3] καὶ νόσον· δῆλον οὐχ ὡς αἱ αὐταί, ἀλλ' ὡς ἕτεραι καὶ ἐναντίαι. θερμὸν δ' ἐπιχέουσιν διὰ τὴν ἀσθένειαν τῶν σπερμάτων[4] (ὅπως μὴ καθάπερ ἐκπαγῇ) διὰ[5] τὸ θᾶττον ἀποδηλοῦν· ἐπεὶ οἵ γε μὴ παντελῶς διεφθαρμένοι καὶ ψυχρῷ[6] διαβλαστάνουσιν· συμβαίνει δὲ καὶ κόπτεσθαι καὶ σήπεσθαι δι' ἀλλοτρίας θερμότητος. [τῶν ἐν τοῖς δένδροις καὶ φυτοῖς τα μεν αυτοματα γίνεται· τα δε εκ παρασκευῆς καὶ θεραπείας][7]

---

[1] P : παρομωνυμ(μ from α U^{cc})ίαν U^{ac} N : παρομωνύμια U^c : παρομοίαν a.
[2] δὲ <γόνιμος> Schneider.
[3] Gaza (faciant), Itali : ποιοῦσα U.
[4] u : σπερμά | U.
[5] <καὶ> διὰ Schneider : διά <τε> Wimmer.
[6] <ἐν> ψυχρῷ Heinsius.
[7] [τῶν — θεραπείας] aP.
subscription: θεοφραστου περι φυτῶν αἰτιων δ̄ U.

appear to be simply a matter of homonymy. For the heat that destroys and the heat that generates differ, and so too does the fluid; and the heat that causes decomposition is alien and destructive, whereas that of the warm water has a certain rightness in its amount, for the seed needs heat of this character for sprouting. Indeed it would be much the same if one should wonder why heat and fluid produce not only health in the body but also disease, for it is evident that they do not do so as the same, but as distinct and opposite. Warm water is poured on the seeds owing to their weakness (to keep them from getting as it were frozen) to make them show more quickly whether they will sprout (in fact wheat that is not completely spoilt sprouts in cold water as well). Both getting worm-eaten and decomposing, on the other hand, are brought about through alien heat.